THE
COLOURS OF

AN INTRODUCTION TO THE CHEMISTRY OF PORPHYRINS AND RELATED COMPOUNDS

Lionel R. Milgrom

Senior Lecturer in Inorganic Chemistry, Brunel University

Oxford   New York   Tokyo
OXFORD UNIVERSITY PRESS
1997

Oxford University Press, Great Clarendon Street, Oxford OX2 6DP

Oxford   New York
Athens   Auckland   Bangkok   Bogota   Bombay   Buenos Aires
Calcutta   Cape Town   Dar es Salaam   Delhi   Florence   Hong Kong
Istanbul   Karachi   Kuala Lumpur   Madras   Madrid   Melbourne
Mexico City   Nairobi   Paris   Singapore   Taipei   Tokyo   Toronto
and associated companies in
Berlin   Ibadan

Oxford is a trade mark of Oxford University Press

Published in the United States
by Oxford University Press Inc., New York

A catalogue record for this book is available from the British Library

Library of Congress Cataloging in Publication Data
Milgrom, Lionel R.
The colours of life : an introduction to the chemistry of
porphyrins and related compounds / Lionel R. Milgrom.
Includes bibliographical references.
1. Porphyrins.   I. Title   QD441.M55   1997   547'.593—dc20   96–27452 CIP
ISBN 0 19 855380 3 (Hbk)
ISBN 0 19 855962 3 (Pbk)

Typeset by Footnote Graphics, Warminster, Wilts
Printed in Great Britain by
Biddles Ltd, Guildford and King's Lynn

PREFACE

Why is grass green? Why not red or any other colour? Come to think of it, why is blood red? Nearly 400 years ago, questions such as these had the best brains of the time (the metaphysicians) scratching their heads: both John Donne and Sir Walter Raleigh thought that our ignorance of the answers to these questions was clear evidence of man's spiritual bankruptcy.

Thanks to modern chemistry, however, we now know that just one family of pigments is responsible for our green world and our red blood. These pigments are based on a macrocyclic ring of carbon and nitrogen atoms, some of the best examples of which are the iron-containing porphyrins found as *heme* (of hemoglobin) and the magnesium-containing reduced porphyrin (or *chlorin*) found in chlorophyll, the pigment that puts the green into plants. Without porphyrins and their relatives, life as we know it would be impossible.

Plants use chlorophyll to collect photonic energy from the sun, which is then harnessed to convert carbon dioxide and water into carbohydrates. Life as we know it depends on this *photosynthetic process*, because plants need carbohydrates to grow, animals eat the plants, and other animals eat them. Quite simply, no chlorophyll, no photosynthesis, no life (except for certain bacteria which get their energy from purely chemical sources). Also, without heme in hemoglobin, large, fast vertebrates could not have evolved, because they would not have had the means of rapidly transporting oxygen around their complex bodies. Their cells, starved of oxygen and energy, would grind to a deadly halt. So where there is life, there are the pigments of life.

But where there *was* life, the fossil remains of these pigments are still to be found. Deep underground, the geological processes which gradually changed dead plant and animal matter into the fossil fuels, coal and oil, also converted the green chlorophyll into red *petroporphyrins*. Geologists use these as chemical markers in their quest for new reserves of oil.

As if to emphasize our complete dependency on these pigments of life, strange disorders called *porphyrias* afflict those with faulty porphyrin metabolism. Such disorders can cause severe light sensitivity and even insanity. King George III may have owed his notorious irascibility and eventual madness to porphyria. There is even the possibility that those mythical denizens of the night, vampires, could have been sufferers of a rare form of this condition.

Finally, porphyrins are being used in the search for new sources of energy, and in attempts to find new ways of curing cancer. The light-absorbing powers of porphyrins and related compounds may be used in the near future to harness the energy of the sun. This same light sensitivity could also make porphyrins more accurate than X-rays at destroying cancer cells. Also, in the near future, porphyrins could figure in an electronics revolution that will produce the carbon-based molecular computer—thousands of times smaller and millions of times more powerful than our present machines.

Clearly, the subject of porphyrins embraces many disciplines, including chemistry, biochemistry, medicine, geology, chemical engineering, paleo-biology, alternative energy, and microelectronics. I have had the privilege of spending twenty years of my life playing with these colourful compounds so that, by way of giving thanks, this book should carry a few dedications.

First, there is my wife, Barbara, and my children, Daniel, Katy, and Amy, who bravely supported me and put up with my late nights (and subsequent bad temper), while writing this book. Then, there are my parents who, in more ways than one, provided me with the tools to do the job. Finally, there are my sadly deceased first and second mentors, Professor George Kenner (who sparked off my interest in porphyrins) and Dr John Dalton (who rekindled the flame). I think the following quote from J.R.R. Tolkien's *Lord of the Rings* sums it up best.

> *All that is gold does not glitter,*
> *Not all those who wonder are lost;*
> *The old that is strong does not wither,*
> *Deep roots are not reached by the frost.*
> *From the ashes a fire shall be woken,*
> *A light from the shadows shall spring;*
> *Renewed shall be blade that was broken,*
> *The crownless again shall be king.*

Uxbridge
December 1996

L.R.M.

CONTENTS

1. What porphyrins are and what they do

1.1 Introduction: the colour purple

What's in a name? Names are important because they can tell us what things are, and what they do. Above all, names tell stories. Take the word *porphyrin* for instance. In a strictly chemical sense, this word has an exact meaning. It is the name given to a family of intensely coloured compounds, at whose heart is a large macrocycle of twenty carbon atoms and four nitrogen atoms. The macrocycle is itself built up from four smaller rings, each made up of four carbon atoms and one nitrogen atom. Each of these rings is joined to its neighbour by one bridging carbon atom. So much for the chemical definition. Hidden within it is a vast chunk of history, stretching back to the beginning of civilisation in the Eastern Mediterranean.

The word *porphyrin* has its origins in the classical world of ancient Greece. In those days, the Greek word *porphura* was used to describe the colour purple.[1] This immediately tells us something about one of the most important features of porphyrins: their intense purple colour. The importance of the colour purple goes back even further in time than ancient Greece. Purple was used as a dye to colour the clothes of royalty and high priests. Old Testament references to royal purple abound. Even now, with monarchies the exception rather than the rule, the colour purple still resonates with wealth, power, and privilege.

The Greek word *porphura* actually derives from an earlier Semitic word used by the Phoenicians to describe molluscs (from the *purpura* and *murex* families) from which they extracted a rare pigment, called Tyrian Purple. Chemically, this substance is known as 6,6'-dibromoindigotin (which is

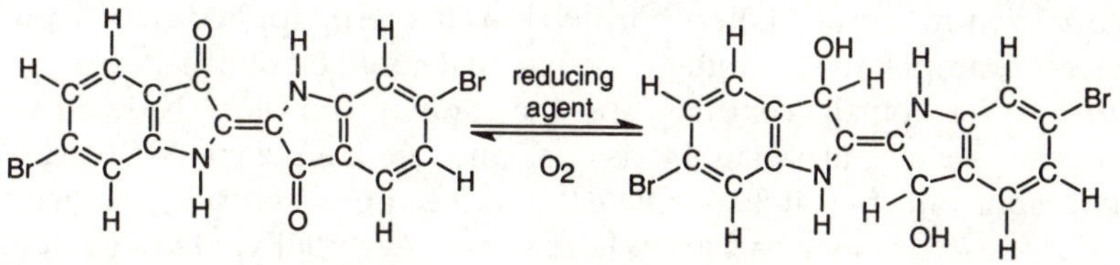

Fig. 1.1 6,6'-Dibromoindigotin and its reduced form. This compound was the basis of the ancient dye Royal Purple, references to which go back to biblical times.

not a porphyrin) and was used to dye the garments of Phoenician royal families.[2] How the Phoenicians actually produced the dye was probably a closely guarded secret, for it is rarely mentioned in any excavated writings. What is known is that there was a large dyeing factory at Sarepta on the Lebanese coast, during the thirteenth century BC.[3] Over a thousand years were to elapse before a detailed account of the ancient Mediterranean dyeing industry was written by Pliny the Elder, in the first century AD.[4] This venerable Roman recounted how the spring was the best time to catch the molluscs, using baited wicker baskets.

Imagine some Phoenician fishermen of 3000 years ago hunting for these rare molluscs, early on a spring morning "just after Sirius has risen in the east". Then, see them later that day, haggling over their precious cargoes at the Tyre dockside. Switch the mind's eye to some dank, dimly lit, fishy smelling room, where the dye was prepared; first by extracting the glands from the unfortunate molluscs, then heating for 10 days in large vats with salt water, periodically skimming the surface for detritus. About ten thousand molluscs would have been required to obtain 1 g of the precious purple, which in some periods would have been worth 10–20 times its weight in gold. The best dyed materials, usually wool, were exceptionally colour-fast (the ancient Greek writer Plutarch wrote that Alexander liberated from the conquered Persian empire, textiles that had been dyed 190 years earlier by the Greeks[5]), and were obtained by double-dipping in the dye. Some variation in the basic purple colour could be obtained by applying human urine. This gave a much redder fabric. Finally, imagine the Phoenician royal court, resplendent in their purple robes. The same Semitic word for the mollusc, eventually came to describe the dye and the cloth coloured with it.

The Greeks continued the tradition of distinguishing their nobility by the colour of their garments. The habit was then taken up enthusiastically in Rome, where the imperial toga was lined with purple. The Emperor Nero was the first to insist that only the ruler had the right to wear royal purple.[2] From classical times, therefore, a link was forged between the colour purple and the idea of noble or royal blood.

The Greek word for purple also crops up in another context. It was used to describe a very hard and expensive purple rock, called porphyry, which was quarried in Egypt.[1] Later, porphyry was a term applied to any purple stone, for example, red granite or marble, that could be polished.

During the fourth century AD, the Roman Empire, bedevilled by internecine strife, barbarian invasions, and an inefficient administrative system, split into two halves. Over the succeeding centuries, the Western half disappeared under barbarian hordes and, eventually, medieval feudalism. The Eastern half, however, transformed itself into the Greek-speaking Byzantine Empire (centred on Constantinople, or Byzantium) which lasted another 1000 years. It was here that the purple/royal connection was further strengthened. The English phrase "to be born into the purple" is, in fact, a literal translation of the Byzantine–Greek word "porphyrogenite".[1] This was a doctrine, first conceived during the reign of the Byzantine Emperor Constantine VII, which stated that the Imperial succession could only pass to the eldest son born *after* the father's accession to the throne. Any children born before this event didn't count—unless, as sometimes happened, they assassinated their father and any inconvenient siblings who stood in their way. As if to highlight this "before and after" difference between Imperial children, Byzantine empresses gave birth to their little porphyrogenites in a special purple room lined with porphyry. To be "born into the purple" was originally meant quite literally, that one was born in the Imperial purple room, built specially for the purpose. Nowadays, it simply means anyone born into a noble or royal family.

1.2 What porphyrins look like

Porphyrins are purple and they are pigments. Such characteristics, however, apply to several dye molecules. What is it that makes porphyrins special? Through heme, an iron porphyrin, they are intimately associated with blood, and many of the redox enzymes involved in metabolic processes. Through chlorophyll, a reduced magnesium porphyrin, they orchestrate photosynthesis, without which life as we know it would be impossible. So what do porphyrins look like?

To expand on the brief definition mentioned earlier, porphyrins are "a large class of deeply coloured red or purple, fluorescent crystalline pigments, of natural or synthetic origin, having in common a substituted aromatic macrocyclic ring consisting of four pyrrole-type residues, linked together by four methine bridging groups."[1]

Pyrrole itself is a five-membered ring of four carbon atoms and one nitrogen atom. To each of the atoms in this ring is also bound a hydrogen atom (Figure 1.2). Imagine the two carbons next to the nitrogen atom, stripped of their hydrogens. To derive the porphyrin macrocycle, we connect four such pyrrole units together via four unsaturated =CH— groups, called methine bridges. The linking together of four pyrrole-like units via methine bridges

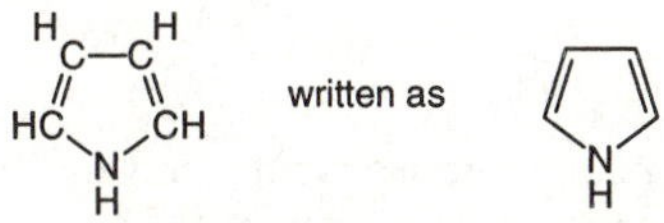

Fig. 1.2 The five-membered heterocycle 1-azacyclopentadiene, better known as pyrrole.

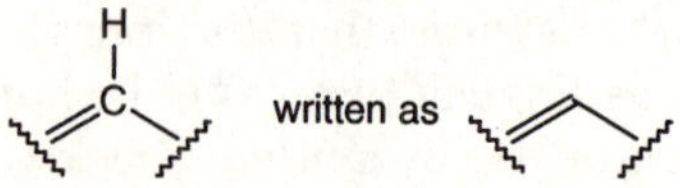

Fig. 1.3 A methine bridge.

produces a macrocyclic molecule whose properties are more than the sum of its parts. Thus, the chemical and physical properties of porphyrins cannot be deduced from those of pyrroles. Pure pyrroles are off-white solids or clear liquids: a far cry from deep purple.

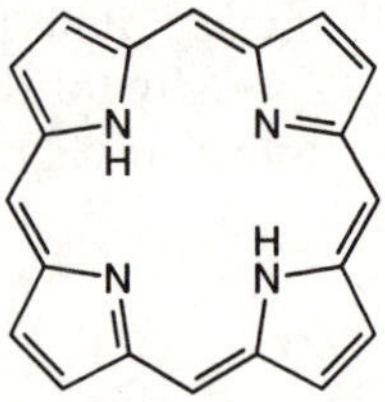

Fig. 1.4 The unsubstituted porphyrin macrocycle.

Interestingly, when the macrocyclic structure of porphyrin was first proposed by Küster in 1912, nobody believed him, least of all Hans Fischer (the father of modern porphyrin chemistry), because such a large ring was thought to be intrinsically unstable. Fischer eventually came round to this structure when he and his Munich school of chemists finally succeeded in synthesising heme—the iron porphyrin in hemoproteins—from pyrrolic starting materials, in 1929.

X-ray crystallography provides stunning confirmation of the porphyrin macrocyclic structure. In the main, this shows that porphyrins are essentially flat molecules. However, circumstances can prevail, when binding

Fig. 1.5 A flat porphyrin, octaethylporphyrin—one of the most popular of the synthetic porphyrins. The large white circles represent nitrogen atoms, black are carbons, and the two tiny white circles are the imino nitrogens. [From D. Dolphin (ed.), *The Porphyrins*, Vol. 3, Academic Press, New York (1978).]

metal atoms, for example, where the macrocycle becomes distinctly buckled with major distortions from planarity. The hole in the middle of the macrocycle can complex a huge variety of metal atoms, with some metals, such as nickel, being just too small to fit into the hole. The macrocycle distorts and twists to maximise its binding to nickel.

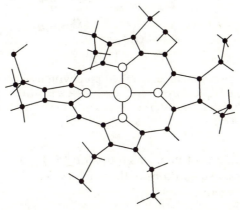

Fig. 1.6 In order to complex a small nickel (largest white circle) cation, the porphyrin macrocycle has to distort. [From D. Dolphin (ed.), *The Porphyrins*, Vol. 3, Academic Press, New York (1978).]

Distortion can occur for other reasons. For example, protonation of the two unprotonated central nitrogens, combined with bulky substituents on the methine bridging carbons, can twist the macrocycle, especially if the bulky substituents attempt to become planar with it. Oxidation of a porphyrin, so that it loses its aromaticity, but retains its integrity as a

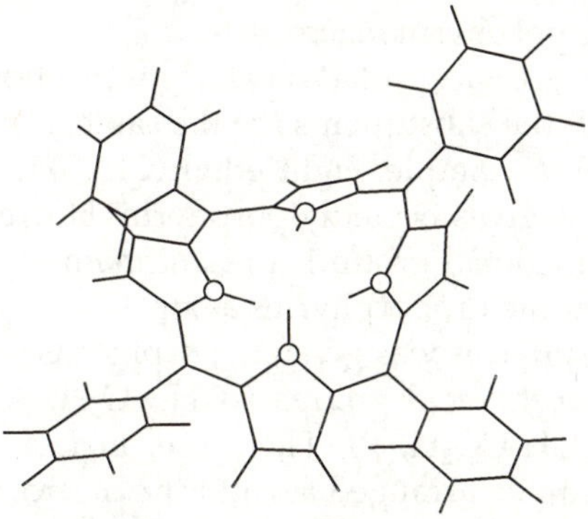

Fig. 1.7 *Meso*-substituted porphyrin dications distort in a different way to accept two more protons. [From D. Dolphin (ed.), *The Porphyrins*, Vol. 3, Academic Press, New York.]

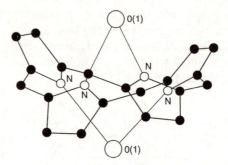

Fig. 1.8 Extreme distortion of an oxidised porphyrin macrocycle. Hydrogen bonds from the central oxygen (large white circles) atoms to the nitrogen atom (small white circles) help hold the distorted macrocycle in shape. (See *J. Chem. Soc., Chem. Commun.*, (1989), 1751.)

macrocycle, and substitution of groups on to the central nitrogen atoms, also lead to massive distortions of the macrocycle.

1.3 Giving porphyrins a good name

It was Hans Fischer who thought up the first system of nomenclature for porphyrins.[6] In line with conventional organic chemical practice at the time, every new porphyrin was given a trivial name. For example, uroporphyrins were so called because they were first isolated from the urine of people suffering from a family of diseases of porphyrin metabolism called porphyrias. As the chemical structures of these newly discovered porphyrins were worked out, the trivial names (when learnt) gave information concerning which organic functional groups were attached to the various carbon atoms of the porphyrin macrocycle.

Many porphyrins are substituted on all eight positions of the four pyrrole fragments. When all the substituents are the same, naming the porphyrin is relatively simple. For example, eight ethyl (CH_3CH_2-) groups attached to the pyrrole positions gives octaethylporphyrin. However, as soon as more than one type of functional group is present, naming becomes much more difficult. Lets look at the uroporphyrins again.

The name uroporphyrins was given to porphyrins with eight substituents consisting of four acetic acid groups ($-CH_2CO_2H$, A) and four propionic acid groups ($-CH_2CH_2CO_2H$, P). There are, in fact, four ways in which these substituents can be arranged around the macrocyclic periphery of the porphyrin. Fischer called these four structures *type isomers*, and denoted them using the Roman numerals, I, II, III, IV. He also called the basic, unsubstituted macrocyclic ring, *porphine*.

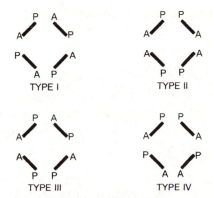

In uroporphyrins there are two types of substituent, $-CH_2.CO_2H$ (A) and $-CH_2.CH_2.CO_2H$ (P). This gives <u>four type</u> isomers depending on which β-positions A and P occupy.

Fig. 1.9 *Four* different type isomers with *two* different β isomers, one of each on each pyrrole ring.

When there are three different substituents, then the number of type isomers rises dramatically. Mesoporphyrins, for example, have four methyl groups (CH_3 Me), two ethyl groups (Et), and two propionic acid groups. There are 15 ways of putting such a combination of substituents together around the macrocyclic periphery (assuming that the methine bridges remain unsubstituted), leading to 15 type isomers (i.e. I – XV). One of these, the type IX isomer, is the substitution pattern possessed by all the naturally occurring porphyrins, and is clearly related to the type III isomer where only two different substituents are present (Figure 1.10). This relationship will become clearer when we consider how nature puts porphyrins together from simple starting materials.

Fischer's semi-trivial nomenclature also included a numbering system for the macrocyclic ring. Each position on the pyrrole fragments where a sub-stituent could go, was called a β-carbon and was given a number 1 to 8. The pyrrole positions next to each nitrogen atom were called α-carbons and remained unnumbered. The bridging carbon atoms were called *meso*-carbons and were given a Greek lower case letter, α – δ. In this way, the

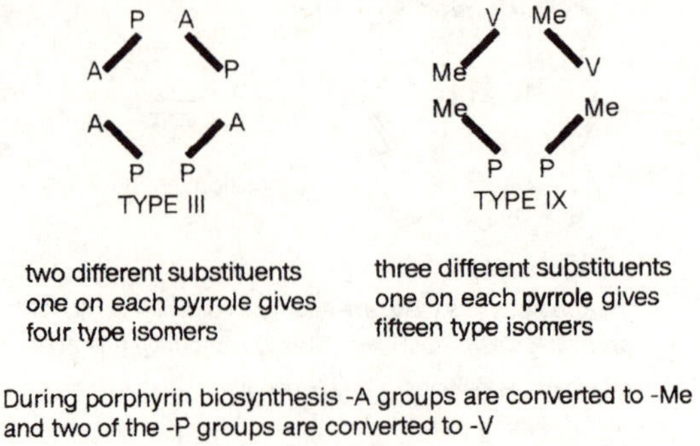

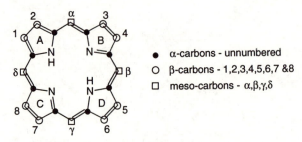

TYPE III

TYPE IX

two different substituents
one on each pyrrole gives
four type isomers

three different substituents
one on each pyrrole gives
fifteen type isomers

During porphyrin biosynthesis -A groups are converted to -Me
and two of the -P groups are converted to -V

Fig. 1.10 Comparison between type III and type IX isomers.

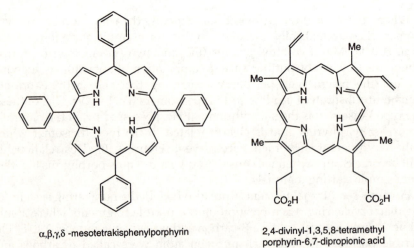

● α-carbons - unnumbered

○ β-carbons - 1,2,3,4,5,6,7 &8

□ meso-carbons - α,β,γ,δ

Fig. 1.11 Fischer's numbering system.

α,β,γ,δ -mesotetrakisphenylporphyrin

2,4-divinyl-1,3,5,8-tetramethyl
porphyrin-6,7-dipropionic acid

Fig. 1.12 Two well-known porphyrins named according to the Fischer nomenclature system.

substitution pattern of a porphyrin was immediately apparent. For example, protoporphyrin IX, the porphyrin present in hemoglobin and many other hemoproteins, would be called *2,4-divinyl-1,3,5,8-tetramethylporphin-6,7-dipropionic acid*. Similarly, a synthetic porphyrin with four phenyl groups substituted in all the *meso*-positions, is called *meso*-α,β,γ,δ-*tetrakisphenylporphin*.

The Fischer system is straightforward for naming simple porphyrins, but as the complexity of a porphyrin derivative increases, with different substituents on the β- and *meso*-positions, the system becomes unwieldy and even contradictory (systems of nomenclature have to be self-consistent).

A new, more systematic IUPAC nomenclature was introduced in 1979 (and finalised in 1987) which numbered all the atoms (including the nitrogen atoms, see Figure 1.13) in the macrocycle, cut down on the number of trivial names used, and also managed to incorporate structural information into the name.[7] The parent macrocycle is called *porphyrin* and the numbering of a substituted porphyrin depends on where on the macrocycle the principle groups (e.g. acetic acid, propionic acid—which appear as suffixes after the word "porphyrin") are present. This helps to place the position of carbon atom number 1, the hydrogen atoms present, and the alphabetical ranking of substituents, considered in ascending order. Thus, what was, mesoporphyrin IX or 2,4-diethyl-1,3,5,8-tetramethylporphin-6,7-dipropionic acid in the old Fischer system becomes *7,12-diethyl-3,8,13,17-tetramethylporphyrin-2,18-dipropionic acid*.

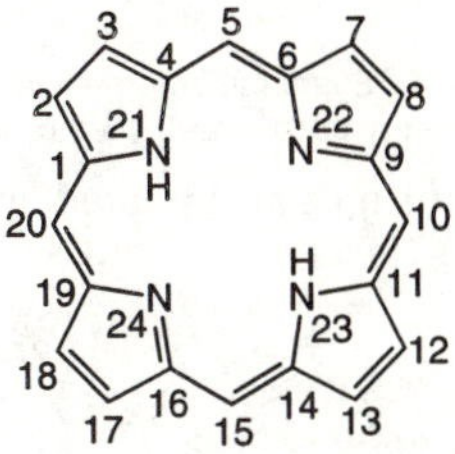

Fig. 1.13 All-number IUPAC systematic nomenclature on macrocycle.

This may seem a senseless change for this porphyrin when compared to the Fischer nomenclature, but in more complicated porphyrins it allows substituents on any of the atoms of the macrocycle to be accurately recorded in the name. So, a substituent on, say, the carbon atom numbered 8, containing a chain of three carbon atoms with a hydroxy group on the second of these carbon atoms, would be numbered so that the first carbon attached to the 8-position is designated as 8^1, while the carbon carrying the hydroxy group is designated 8^2. In addition, cutting down on the number of trivial names, shows the familial relationship between different porphyrin structures. Fortunately, the IUPAC committee in charge of porphyrin nomenclature realised this. So, alongside the systematic (and sometimes

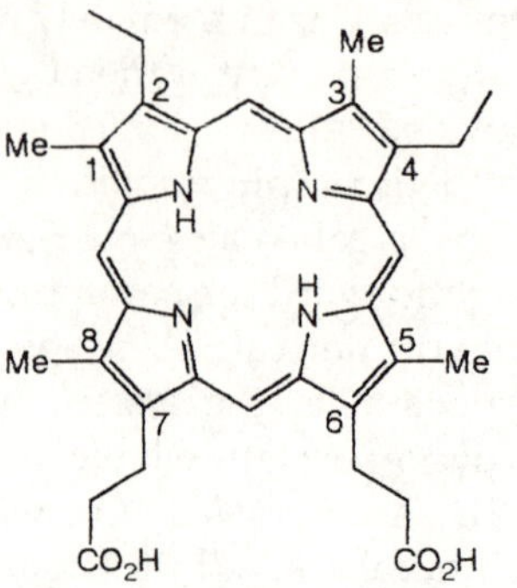

Mesoporphyrin IX - Fischer nomenclature; 2,4-diethyl-
1,3,5,8-tetramethylporphyrin-6,7-dipropionic acid

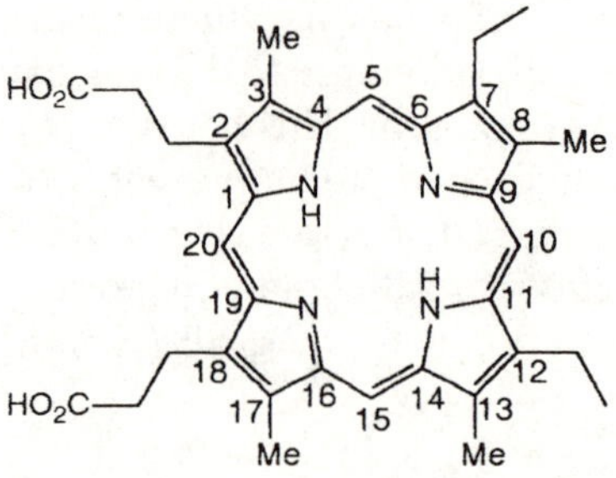

Mesoporphyrin IX - Systematic nomenclature; 7,12-diethyl-
3,8,13,17-tetramethylporphyrin-2,18-dipropionic acid

Fig. 1.14 Mesoporphyrin IX and related systematic numbering.

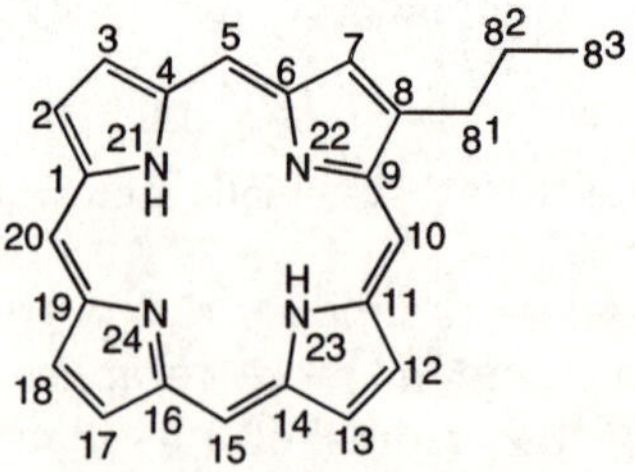

Fig. 1.15 All-number nomenclature of porphyrin macrocycle, plus numbering of a side chain on position 8.

incredibly long-winded) IUPAC nomenclature, there is also a semi-systematic nomenclature based on a dozen trivial names. These are ranked in order of importance starting with deuteroporphyrin IX, now just called deuteroporphyrin.

When there is a choice of trivial names, the one with the highest ranking

Trivial Name	Rank	Substituents and locants								
		2	3	7	8	12	13	15	17	18
Coproporphyrin I	9	Me	Cet*	Me	Cet	Me	Cet	H	Me	Cet
Cytoporphyrin	11	Me	−CH(OH)CH₂R′	Me	Vn	Me	Cet	H	Cet	−CHO
Deuteroporphyrin	1	Me	H	Me	H	Me	Cet	H	Cet	Me
Etioporphyrin I	3	Me	Et	Me	Et	Me	Et	H	Me	Et
Hematoporphyrin	8	Me	−CH(OH)CH₃	Me	−CH(OH)CH₃	Me	Cet	H	Cet	Me
Mesoporphyrin	7	Me	Et	Me	Et	Me	Cet	H	Cet	Me
Phylloporphyrin	4	Me	Et	Me	Et	Me	H	Me	Cet	Me
Protoporphyrin	6	Me	Vn	Me	Vn	Me	Cet	H	Cet	Me
Pyrroporphyrin	2	Me	Et	Me	Et	Me	H	H	Cet	Me
Rhodoporphyrin	5	Me	Et	Me	Et	Me	−CO₂H	H	Cet	Me
Tropoporphyrin I	10	Cm	Cet	Cm	Cet	Cm	Cet	H	Cm	Cet
Phytoporphyrin	12	Me	Et	Me	Et	Me −C(O)−CH₂−			Cet	Me

* Cet = −CH₂.CH₂.CO₂H.

Fig. 1.16 Ranking of twelve trivial porphyrin names.

in the table is preferred. Prefixes and/or suffixes of substituents are usually combined with the trivial name. For example, the porphyrin shown in Figure 1.17, which is found as its metal complex in oil shales and other fossil fuel deposits, is called deoxophylloerythroetioporphyrin (DPEP) in the old Fischer system. In the revised nomenclature system, all porphyrins with exocyclic rings are related to phytoporphyrin (Figure 1.16). Compared with this, DPEP lacks a carboxylic acid group at the end of the propionic

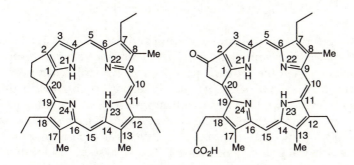

Fischer: Deoxophylloerythroetioporphrin

Systematic: 7,12,18-triethyl-2¹,2²-dihydro-3,8,13,17-tetramethylcyclopenta[a,t]porphyrin

Trivial name: phytoporphyrin

Systematic: 7,12-triethyl-2²-hydro-3,8,13,17-tetramethyl-2¹-oxocyclopenta[a,t]porphyrin-18-propionic acid

Fig. 1.17 The different nomenclatures for two related porphyrins.

acid residue on position 17 (i.e. on the 17^2-position), and an oxygen atom on the first carbon atom of the exocyclic ring (i.e. on the 13^1-position). This lack of the required groups to make a full phytoporphyrin is recorded in the name as *17^2-decarboxy-13^1-deoxophytoporphyrin*.

There are many variations on the porphyrin macrocycle, each with its own name. For example, if the double bond between carbons 17 and 18 is reduced (so that these carbon atoms are fully saturated) we have a *chlorin*. Such a reduced porphyrin macrocycle is to be found in chlorophyll. Similarly, if the double bond between carbons 7 and 8 is also reduced (so that two of the original porphyrin double bonds have been taken out) then we have a *bacteriochlorin*. In fact, there are two ways in which two double bonds could be removed to give bacteriochlorins—from pyrroles either opposite or adjacent to each other. These are shown in Figure 1.18.

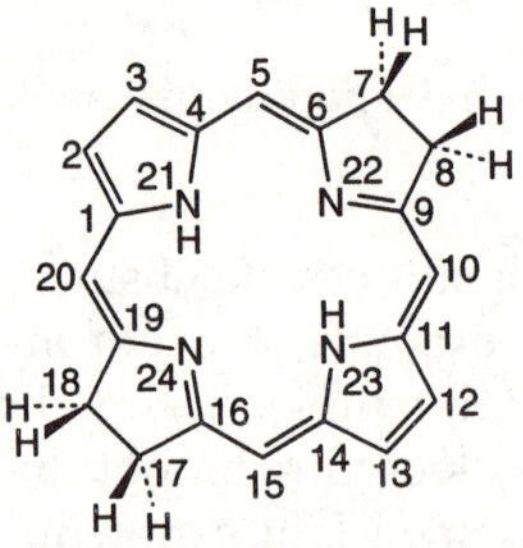

Bacteriochlorin
or
7,8,17,18-tetrahydroporphyrin

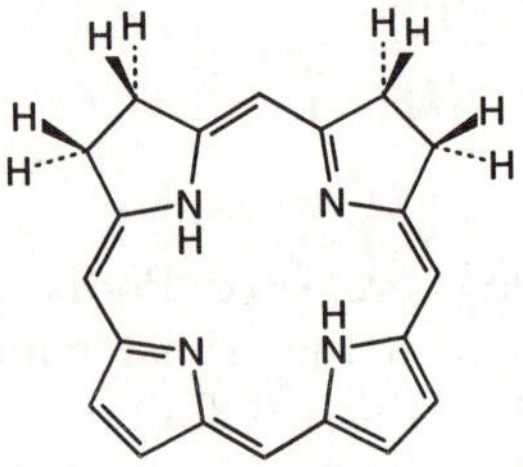

Isobacteriochlorin
or
2,3,7,8-tetrahydroporphyrin

Fig. 1.18 Two types of bacteriochlorin.

IUPAC non-trivial names for the chlorin and bacteriochlorin macrocycles are *17,18-dihydroporphyrin* and *7,8,17,18-tetrahydroporphyrin*, respectively. (The situation when the reduced pyrrole rings are adjacent to each other would be *2,3,17,18-tetrahydroporphyrin*). Referring to chlorins in this way shows their isomeric relationship with macrocycles at the same oxidation level that are reduced at one of the *meso*-positions, e.g. the *phlorins* or *5,22-dihydroporphyrins*. Because phlorins are reduced at a position which interrupts the delocalisation pathway of the π-system, phlorins readily isomerise into the fully delocalised (and therefore more stable) chlorins.

Another important natural variation on the porphyrin theme are the ring-contracted *corrins*, the most famous example being the cobalamin macrocycle vitamin B_{12} coenzyme. Here, one of the *meso*-carbons is missing.

There are other types of modified porphyrin macrocycle which will be mentioned when the need arises. However, no survey of porphyrin structural types, however brief, could afford to ignore arguably the most useful

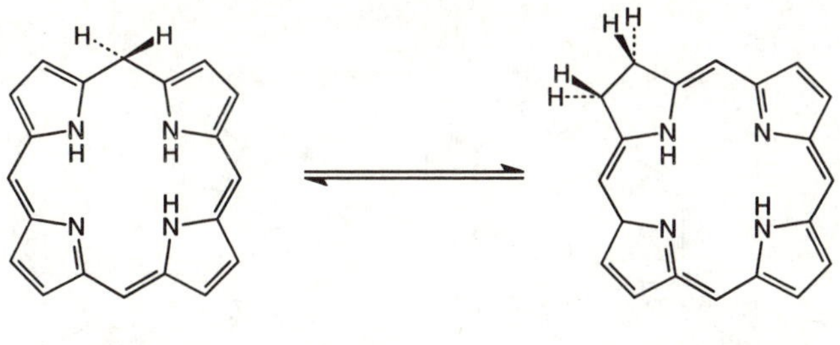

phlorin
or
5,22-dihydroporphyrin

chlorin
or
2,3-dihydroporphyrin

Fig. 1.19 Tautomerism of a phlorin to a chlorin. In the former, the delocalisation pathway is broken.

porphyrin analogues yet discovered, the *phthalocyanines* (or *5,10,15,20-tetraazatetrabenzoporphyrins*). These were discovered purely by chance in a Scottish chemical production plant in 1928. They are dark blue and highly stable, and their potential as dyestuffs and pigments was quickly recognised.

1.4 What porphyrins do

1.4.1 Introduction

One of the more interesting things about porphyrins is how small variations on the basic structural theme of a tetrapyrrolic macrocycle, leads to a wide diversity of biochemical functions. Looked at side by side, it may seem whimsical to think that heme and chlorophyll are related molecules. However, up to a point they share the same biosynthetic pathways. Only near the end of these paths do their biosyntheses diverge. If the molecules are examined closely enough, the broad outline of their family relationship can be discerned—it's like looking at first cousins in an album of family photographs. Even vitamin B_{12}, the macrocycle of which is the profoundly modified an shrunken corrin ring system, is still recognisable as family. The similarity between heme and chlorophyll is no coincidence but points to an underlying economy in the way these biopigments were designed, and is evidence of the common ancestry of all living things.[8]

The main function of porphyrins and porphyrin-like compounds in nature is to bind metal atoms, which act as centres for significant biochemical events. Thus, protoporphyrin IX in heme, complexes iron which, in hemoglobin and myoglobin, reversibly binds oxygen so that it can be transported around the body (hemoglobin) or stored in muscle tissue (myoglobin). The fascinating thing here for inorganic chemists is that, unlike in

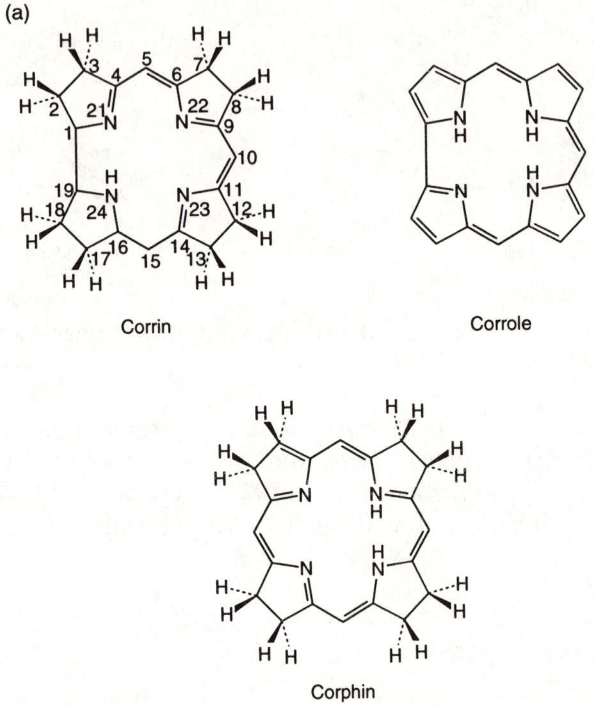

(a)

Corrin

Corrole

Corphin

Fig. 1.20 (a) Corrin, corrole, and corphin macrocycles. Notice how C-20 is missing and the numbering of the macrocycle reflects this. (b) Coenzyme vitamin B_{12}.

its native state, the iron does not irreversibly change its oxidation state upon binding oxygen. The blend of porphyrin ligand and protein environment changes the redox potential of the iron and stops the latter complexing with water. Taken together, these factors inhibit the usual oxidation of Fe(II) to Fe(III) in the presence of air and water.

Changing the external protein environment of the heme (which partly changes the way the iron atom is bound) leads to completely different iron chemistry. Thus, in cytochrome *c*, the iron atom cycles through the +2 and +3 oxidation states, while performing the function of electron transfer in cell respiration. Thus, with a subtle blend of thermodynamics and kinetics, nature coaxes iron into doing its biochemical bidding.

In vitamin B_{12} coenzyme, the only naturally occurring organometallic compound (i.e. containing a metal–carbon bond), nature makes use of cobalt to reduce organic species and in reactions involving the transfer of hydrogen atoms. Lack of this vitamin leads to the deficiency disease pernicious anaemia in which there is a shortage of red cells and hemoglobin, and impairment of the central nervous system. This is rarely caused by a dietary deficiency and is mainly a result of imperfect absorption of the vitamin by the intestine.

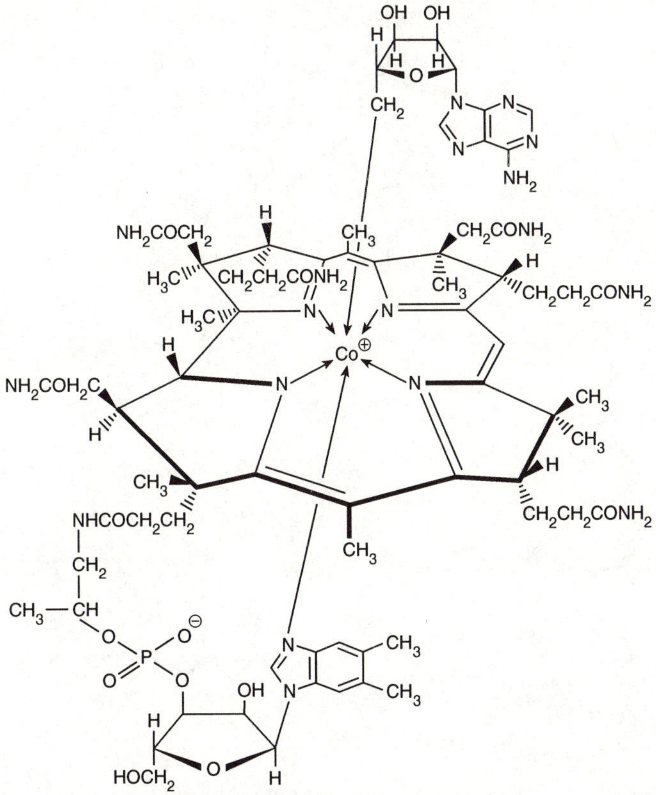

Thus, the accessibility of porphyrins and related macrocyclic rings is one of the factors that allows biological systems to capitalise on and to modify for their own purposes, the redox and coordination properties of metal ions.

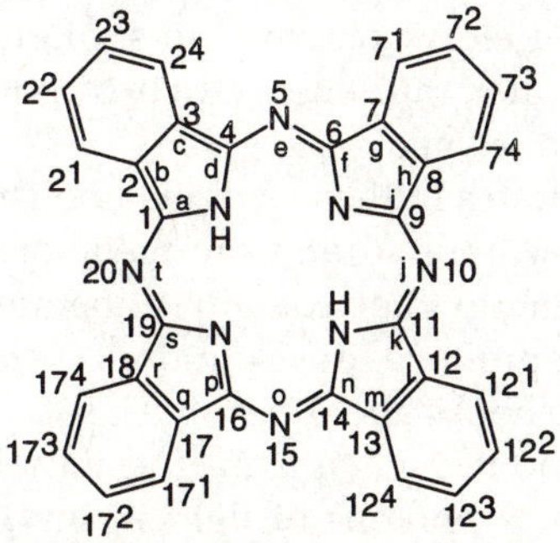

Fig. 1.21 Phthalocyanine or tetrabenzo[b,g,l,q]-5,10,15,20-tetraazaporphyrin.

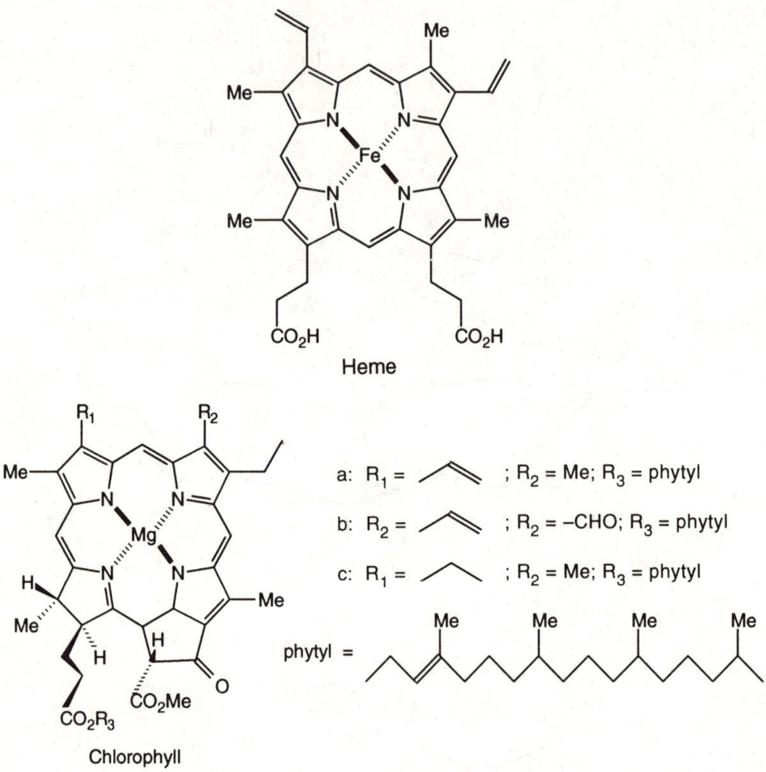

Fig. 1.22 Heme and some chlorophylls.

1.4.2 Photosynthesis

Without photosynthesis, there could be no life as we know it. Not only does photosynthesis supply the oxygen we breathe (as a mere by-product), it also provides most of the energy that drives living things. Photosynthesis does this by trapping energy contained in sunlight and using it to convert carbon dioxide and water into energy-rich carbohydrates. In the process, the water is oxidised to oxygen.

Plants use carbohydrates in their own metabolism, as do the animals that eat the plants. As the carbohydrates are metabolised back into carbon dioxide and water, the energy they contain is liberated and used by an organism to grow, move, reproduce, devise more efficient ways of killing other organisms—and write books about porphyrins.

In chlorophyll, the bound metal is magnesium. Here, the function of the macrocycle is to capture photons of light in the near-ultraviolet (400 nm) and red (650–700 nm) regions of the visible spectrum. In the reduced por-

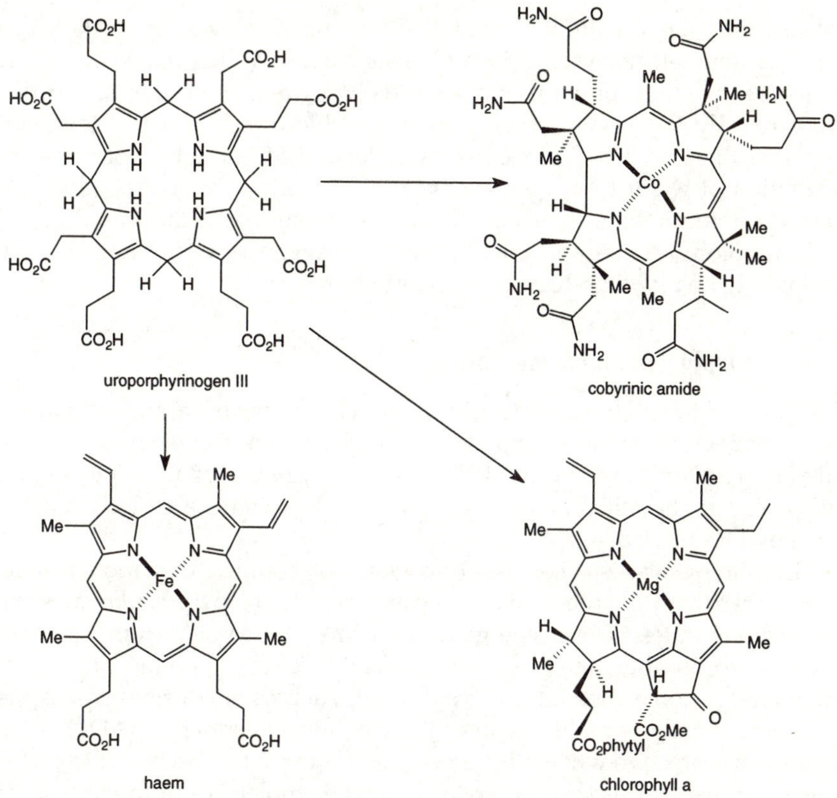

Fig. 1.23 Family relationships between heme, chlorophyll *a*, and cobyrinic amide, the heart of vitamin B_{12}. They all share a common biosynthetic precursor, uroporphyrinogen III.

phyrin macrocycle of chlorophyll, the conjugated system of double bonds is ideally suited to this task, while the substituents around the macrocycle serve to fine tune its light-absorption characteristics. The metal also serves to modulate the light-absorbing and energy-transfer characteristics of the chlorophyll, while acting as a centre for binding of water (needed as a source of electrons for the photosynthetic process).

In fact, chlorophyll molecules do not work in isolation. Rather, batches of 2–300 chlorophyll molecules are arranged into tiny molecular antennae that harvest large quantities of photons in a short space of time and focus them on to a "special pair" of chlorophyll molecules. Here, the energy of the photons is trapped as excited electrons, which are led away by a chain of electron-transporting proteins to begin the work of generating the high-energy molecules necessary for turning carbon dioxide into carbohydrates.

Meanwhile, the stream of excited electrons flowing away from the chlorophyll antenna-trap system, is replenished from water molecules coordinated to the chlorophyll. In other words, the water molecules are oxidised very rapidly to oxygen. Having so many chlorophyll molecules gathered together ensures rapid and efficient collection of enough photons so that oxidation of water to oxygen occurs rapidly with very little production of harmful intermediates (such as peroxide or superoxide) that could damage delicate biomolecules and membranes. A manganese-containing protein is intimately involved in the water-oxidation process.

1.4.3 Hemoglobin and hemoproteins

Without oxygen, we suffocate and die. What that means at the cellular level is that metabolism stops. So our dependence on the oxygen-producing higher green organisms is total. But why is oxygen so vital? What is it actually doing in the cell? Put simply, oxygen acts as a sink, a molecular collecting bowl for electrons.

Cellular metabolism consists of a chain of coupled oxidation reactions which start with the removal of electrons from high-energy substrates, such as carbohydrates. These energetic electrons are passed from one redox enzyme to another, their energy gradually being syphoned off to make high-energy molecules for use by the cell, such as adenosine triphosphate (ATP) and nicotinamide adenine dinucleotide phosphate (NADP). At the end of this process, when all the necessary energy has been extracted from the electrons, they are carried away by oxygen molecules (four electrons per oxygen molecule), which, after each one picks up four protons, are each converted into two harmless water molecules. Oxygen, therefore, removes some of the "ash" from the metabolic "burning" of carbohydrates. It is therefore very much in the interests of a higher organism to get oxygen to its different parts as rapidly and as efficiently as possible (small, single-cell organisms do not require an oxygen-transport system beyond simple diffusion), to store oxygen (especially in highly energetic tissues like muscle cells), and to remove the other major constituent of metabolic "ash", i.e. carbon dioxide. These functions are fulfilled by hemoglobin and myoglobin, the carrier and storer of oxygen, respectively.

Hemoglobin transports oxygen from its source in the organism, e.g. lungs, gills, or skin, to the site where it is needed, mainly muscle tissue. Hemoglobin also carries carbon dioxide away, back to where it can be easily expelled. Just how this is done forms the subject of a later chapter. The oxygen is transferred to myoglobin for storage, ready for use in metabolism. The latter involves a sequence of different hemoproteins (i.e. proteins containing heme, iron(II) protoporphyrin XI, or its derivatives) called cytochromes whose redox potentials are tuned so that electron flow is, energetically speaking, downhill, finishing with oxygen. The interesting thing here is that the hemoproteins, hemoglobin, myoglobin, and the cytochromes, all share a similar prosthetic group, based on heme. The way this group functions, be it for oxygen transport, storage, or electron transfer, is tailored by the substituents around the porphyrin ring, and the other ligands bound to the central iron moiety. The latter, in turn, are governed by the protein. On top of this, the protein provides a variety of micro-environments which in hemoglobin and myoglobin, for example, bury the heme deep within a hydrophobic interior, so hindering entry of anything other than small gas molecules, such as oxygen. Because water is excluded, the usual iron redox chemistry (which leads to rust in the case of unbound iron) does not happen (water is necessary for electron transfer to take place). In addition, the other ligands present (an imidazole moiety above the iron and one below but further away) also raise the redox potential of the heme, another disincentive for electron transfer to take place. The heme-bound iron in hemoglobin thus remains in its oxidation state of II.

In cytochrome c, on the other hand, the heme is closer to the surface of the protein. Also, the other ligands are bound so strongly that oxygen cannot coordinate. This microenvironment favours electron transfer, so that the iron cycles *reversibly* through its II and III oxidation states.

At this stage, we can begin to appreciate just how subtle a chemist nature is. For heme, (i.e. iron(II) protoporphyrin IX) undergoes an irreversible reaction in the presence of oxygen on its own. Two heme units come together to form a molecular dimer in which oxygen bridges the two iron atoms in their III oxidation state, Fe(III)–O–Fe(III). Such a state of affairs would be disastrous for living processes.

We cannot leave this introduction to what porphyrins do, without mentioning some of the other important hemoproteins, e.g. the dismutases, peroxidases, and catalases. They protect the body against the ravages of oxygen. It may seem strange to think of oxygen as dangerous because, without it, we die. However, the danger in oxygen lies precisely in its role as a scavenger of electrons. Although when oxygen picks up electrons and protons, it is finally reduced to a harmless product, water, it is the intermediates that are potentially dangerous. Intermediates such as superoxide, O_2^- and peroxide, O_2^{2-} are powerful oxidising agents which can play havoc with living tissue. This is where superoxide dismutase, peroxidase, and catalase come in. They act as one line of defence for an organism against the dark side of oxygen. Superoxide dismutase converts superoxide into peroxide while peroxidase destroys peroxides by transferring oxygen to a sacrificial oxidisable substrate. Catalase breaks down hydrogen peroxide before it has a chance to build up in the system.

Hemoproteins are also involved in gene regulation, iron metabolism (cytochrome P450), drug metabolism, and hormone synthesis. Truly, where there is life there are porphyrins.

1.5 How to draw a porphyrin

Throughout this book, you will see many porphyrin structures depicted. If the printer has got it right, then these structures will look immaculate and geometrically perfect. But if you have bought this book as an adjunct to a study of macrocyclic chemistry (you might even be lucky enough to have a whole course in porphyrin chemistry itself), biochemistry, or bio-organic chemistry, then you will need to be able to draw the porphyrin macrocycle rapidly and confidently during a lecture.

Many people make the mistake of trying to draw the macrocyclic outline, hoping that in the fullness of time their eye and hand will become accustomed to rapid, precise sketching. In reality, what happens is that panic sets in as the lecturer races ahead, while the poor student is stuck drawing the same porphyrin. So here is the way to do it, as taught to me by my colleagues in (the late) Prof. G.W. Kenner's research group at Liverpool University in 1969.

Step 1. Set up a square of four nitrogen atoms.

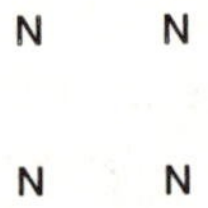

N   N

N   N

Fig. 1.24 First, set up a square of four nitrogens.

Step 2. Construct the inner macrocyclic core of α-carbons and *meso*-carbons.

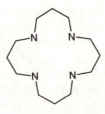

Fig. 1.25 . . . then draw in the inner 16-tetrazaannulene.

Step 3. Fill in the eight β-carbons to give the four pyrrole units embedded in the inner macrocyclic core. This is the basic macrocycle.

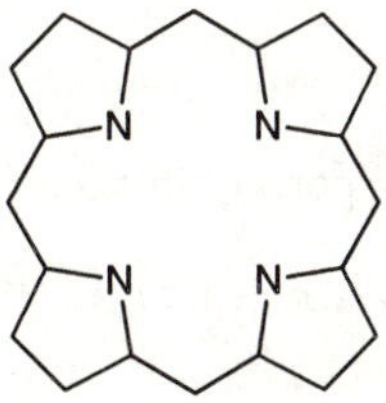

Fig. 1.26 ... add the outer pyrrole rings.

Step 4. Now for the double bonds. Remember, two of the pyrrole units are pyrrolic, with hydrogen atoms on their nitrogens, while the other two are pyrrolenic, which means their nitrogens have no hydrogen attached and are akin to the nitrogens in pyridine.

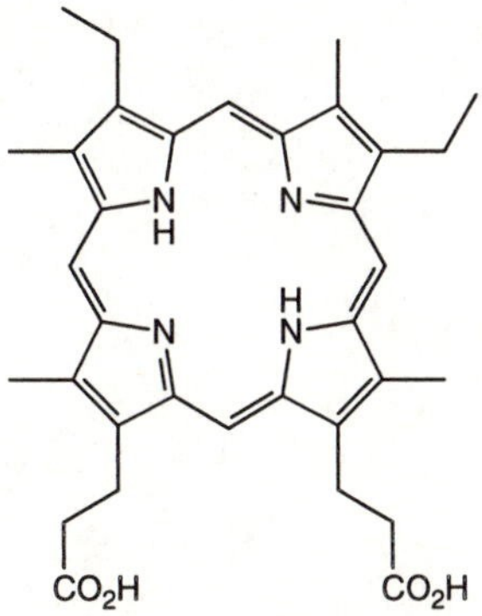

Fig. 1.27 Finally, add double bonds, inner hydrogens (or metal), and outer substituents.

Step 5. The final step is to put in whatever substituents are present on the macrocycle, exocyclic rings (chlorophylls), etc.

1.6 References

1. *The Oxford English Dictionary*, Vol. XII, 2nd Edn, Clarendon Press, Oxford (1989), pp. 138–140.
2. P.E. McGovern and R.H. Michel, *Acc. Chem. Res.*, (1990), **23**, 152.
3. J.B. Pritchard, in *Recovering Sarepta, A Phoenician City*, Princeton University, New Jersey (1978).

4. Pliny the Elder, *Pliny's Natural History*, translated by J. Bostock and H.T. Riley, H.G. Bohn, London (1855).
5. Plutarch, *Lives of the Noble Greeks*, ed. E. Fuller, N. Doubleday, New York (1959), pp. 240–314.
6. H. Fischer and H. Orth, *Die Chemie des Pyrrols*, Vols I–III, Akad. Verlagsges., Leipzig (1934–1940).
7. For a thorough exposition of porphyrin nomenclature, see, *Pure & Appl. Chem.*, (1987), **59**, 779.
8. A. Eschenmoser, *Angew. Chem., Int. Ed. Engl.*, (1988), **27**, 5.

2. Where porphyrins come from . . .

2.1 Earth before life

It is difficult to imagine the earth as a young world devoid of all life. The famous Walt Disney cartoon *Fantasia* gives as good a picture as any of what our prebiotic planet must have been like. The scientific imagination has managed to add some bones to Hollywood's speculations, in the form of experiments on mixtures of gases that are thought to have constituted the earth's primordial atmosphere.

The primoridal scene is imagined as one of huge geological violence. The young earth probably condensed from the same vast cloud of dust and gases as our sun and solar system. Many such dust clouds have been observed within our own galaxy. According to the Kant–Laplace theory, to which most modern theories of stellar and planetary evolution can be traced, such a cloud can condense under the influence of its own gravitational attraction.[1,2b] If the cloud is also slowly spinning, then it will gradually flatten into a disc. These theories do have their problems, however, as they have yet to explain why most planets rotate about their axes with a period of about a day or less, and why the sun, with 99.9% of the mass of the solar system, contains only 2% of its angular momentum.

If the cloud is large enough, then its gravitational field will begin to make gas and dust accumulate in the centre of the disc. The gravitational force continues to pack material into the centre so that the temperature begins to increase. (This is similar to what happens if you place your finger over the end of a bicycle pump and ram home the plunger—the air cannot escape, is compressed, and so warms up.) Eventually, if the mass of material condensing is large enough, the inward gravitational pressure causes the temperature to climb so high that it triggers a thermonuclear reaction in the centre of the cloud. A star is born, and the resulting radiation pressure outwards balances the gravitational pressure inwards.

Meanwhile, the contraction of the cloud causes its rate of rotation to speed up (rotate a conker on the end of a string—as you shorten the string, the rate of rotation increases). This, so the theory goes, produces a tendency for some of the condensing material to be thrown back into space. So not all of it would condense to form the star. This uncondensed material would remain in regions far from the star and would probably take part in separate condensations, sweeping up solid dust particles and gas left in the shrinking dust cloud. This, however, does not adequately describe how the almost mathematically exact spacings between the planets (known as Bode's Law) came about.

One of these secondary condensations resulted in the appearance of the earth some 4.5 billion years ago, according to two independent assessments of the earth's age (meteorite age determination and measurement of lead isotope ratios from terrestrial rocks[2a]). What condition the earth would have been in is a matter of conjecture. Was it hot or cold? Arguments have been put forward to support either condition. But it appears that earlier in its history, the planet was subject to greater volcanic activity than it is now.

Vulcanism is believed to be sustained by the decay of radioactive substances deep within the earth. As the amount of these substances was bound to have been greater 4.5 billion years ago (and as such substances, being heavy, would be likely to be part of the core around which a planet might condense), it seems reasonable to suppose that Walt Disney was right and that volcanic activity provided the geological accompaniment to the birth of the planet. Besides, the amount of cosmic debris left over from the formation of the planets 4.5 billion years ago would have been enough to sustain a greater number of meteoric impacts (with their concomitant release of huge amounts of energy as heat) than is, fortunately, suffered by the earth now. Increased vulcanism would have led to a high degree of outgassing, thus generating the earth's primeval atmosphere. Before we speculate on what the constituents of this atmosphere might have been, let us consider the earth's present atmospheric constituents.

The earth's atmosphere contains approximately 80% free nitrogen and 20% free oxygen. Trace amounts of other substances exist, including carbon dioxide, the noble or rare gases (neon, argon, krypton, and xenon), and oxides of nitrogen and sulphur. But the significant feature is the large amount of free oxygen present. Such an atmosphere is strongly oxidising and would oxidise any organic molecule released into it to carbon dioxide.[3]

For a long time, the earth has possessed a highly oxidising atmosphere that could not have been produced without biotic help. If such an atmosphere had been produced right at the beginning of the earth's history, within a short space of time there would have been no free oxygen left, as it would have quickly combined with everything that it could. It is possible for some oxygen to appear abiotically from the action of ultraviolet light on water. The latter would have been in the atmosphere from volcanic outgassing. However, the amount of oxygen produced (and its consumption during oxidation reactions) would still not account for the large percentage currently extant. This is due entirely to photosynthesis by algae and land plants.

It is generally agreed that the early atmosphere was a reducing one and most likely consisted of water, methane, carbon dioxide, and nitrogen, probably with traces of ammonia and hydrogen thrown in for good measure. Such a mixture has been shown to generate some of the simple organic molecules, necessary as basic building blocks for life, when subjected to electric discharges (the laboratory equivalent of lightning flashes) and ultraviolet light.[4] It is also likely that sound, in the form of concussions from lightning flashes and geological eruptions due to vulcanism, played an important part in promoting chemical reactions in the primordial atmosphere.

So, how would the prebiotic synthesis of molecules have occurred and what molecules would most likely have been produced? To an early nineteenth century chemist, such a question would have seemed very strange. We still divide the science of chemistry into two parts; organic and inorganic. In the old days, the former dealt with the chemistry of molecules containing carbon that were derived from living organisms of some kind, while the latter dealt with every other element in the periodic table and their compounds, which were considered lifeless. Somehow, it was imagined, there was an unbridgeable difference between organic and inorganic molecules. Then, in 1832, Friedrich Wöhler heated ammonium cyanate (an inorganic compound) and converted it into urea (a well-known organic species). Since then, many such syntheses of organic compounds from inorganic starting materials have been accomplished.

For example, the simple amino acid glycine can be made from hydrogen cyanide, HCN. Such experiments could be said to constitute early attempts at prebiotic chemistry. It wasn't until 1953, however, when Stanley Miller, then a Harvard postgraduate student, subjected a reducing "atmosphere" of methane, ammonia, hydrogen, and water to an electric discharge for a week.[4] Miller tried to mimic what he thought were some aspects of the prebiotic earth, such as lightning, while boiling water represented the primeval "ocean". After a week, Miller found that 15% of the carbon that had been in the atmosphere was now dissolved in the water as organic compounds. There was also a significant quantity of tar, some cyanide, and aldehyde (see Figure 2.1).

Since Miller's time, most of the important amino acids have been identified in the prebiotic laboratory "soup", as well as sugars and nucleotide bases. Interestingly, similar experiments performed in the presence of free oxygen, or just carbon dioxide, nitrogen, and water (i.e. oxidising atmospheres) gave no amino acids. On this basis, prebiotic chemists have argued that the primeval atmosphere must have been reducing, otherwise the molecules necessary to get life started could not have been formed.

The mechanism by which amino acids are made seems to involve the prior formation of hydrogen cyanide and aldehydes, which then react with ammonia and water to form amino acids. It is possible to synthesise practically the whole range of amino acids necessary for life by adding small

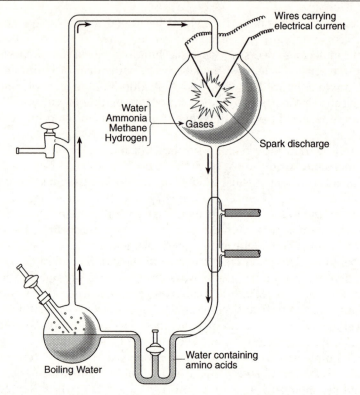

Fig. 2.1 Apparatus used in Miller's experiments on abiotic synthesis of organic molecules. [From L.E. Orgel, *The Origins of Life: Molecules and Natural Selection*, Chapman & Hall, London (1973), p. 126.]

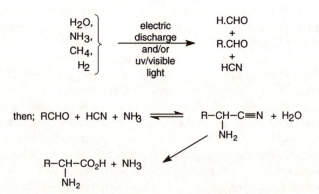

Fig. 2.2 One possible scenario for prebiotic amino acid synthesis. For sulphur-containing amino acids, H_2S and CH_3SH can be added.

amounts of other constituents to the "soup", e.g. sulphur-containing compounds to generate cysteine and methionine. A certain amount of sophistication is added to these experiments when solid particles are introduced into the reaction. These would act as supports and surfaces catalysing many of the reactions and encouraging them to occur under less reducing conditions.[5]

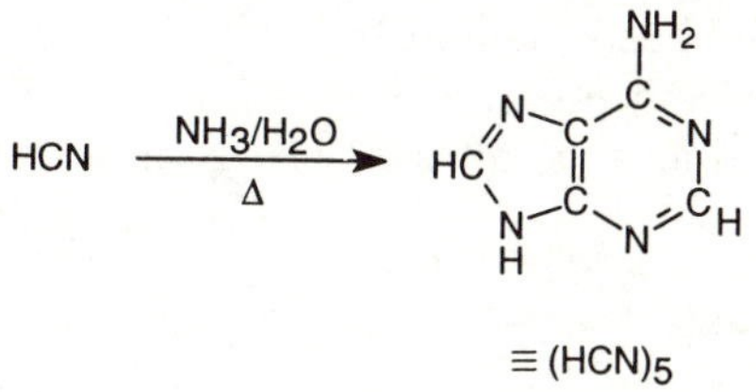

$\equiv$ (HCN)$_5$

Fig. 2.3 Adenine will self-assemble from HCN and may be considered its pentamer.

Amino acids were not the only organic compounds to appear in the soup. Other workers besides Miller were investigating prebiotic synthesis. Oro found that solutions of ammonia and cyanide would produce adenine, one of the organic bases important in nucleic acid synthesis. Adenine belongs to the purine family of organic bases, and may be considered a pentamer of hydrogen cyanide. Other components of DNA can also be made by prebiotic synthesis, including the pyrimidine family of organic bases, and sugars. The former, once again, come from hydrogen cyanide, while the latter can be traced back to formaldehyde.

So far, all the prime constituents of nucleic acid synthesis have proved surprisingly easy to prepare by prebiotic synthesis. What has proved difficult has been the synthesis of nucleosides (organic bases and sugars

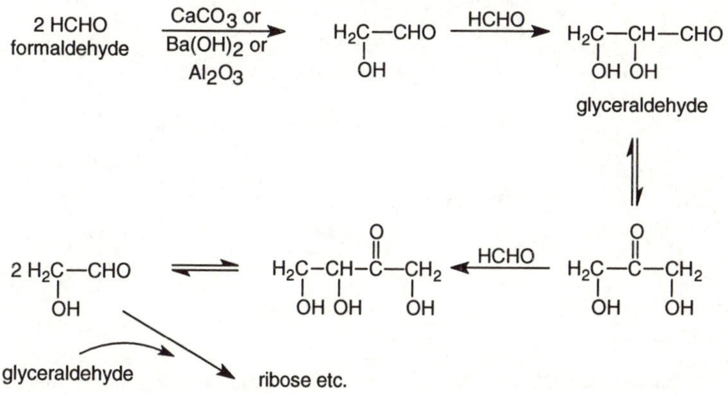

Fig. 2.4 Possible prebiotic formation of sugars from formaldehyde.

combined) and nucleotides (nucleosides plus inorganic phosphate residues), the basic building blocks of nucleic acids.

There have also been claims for a prebiotic synthesis of porphyrins. An electric discharge applied to a gaseous mixture of methane, ammonia, and water gives large yields of formaldehyde, small amounts of pyrrole, and tiny amounts of porphyrin. More efficient syntheses of porphyrins, especially those related to protoporphyrins and uroporphyrins, seem likely. One possibility has been demonstrated by Albert Eschenmoser, who managed to synthesise a mixture of porphyrins in the absence of oxygen and water, from α-amino nitriles using montmorillonite clay as a catalyst.[6]

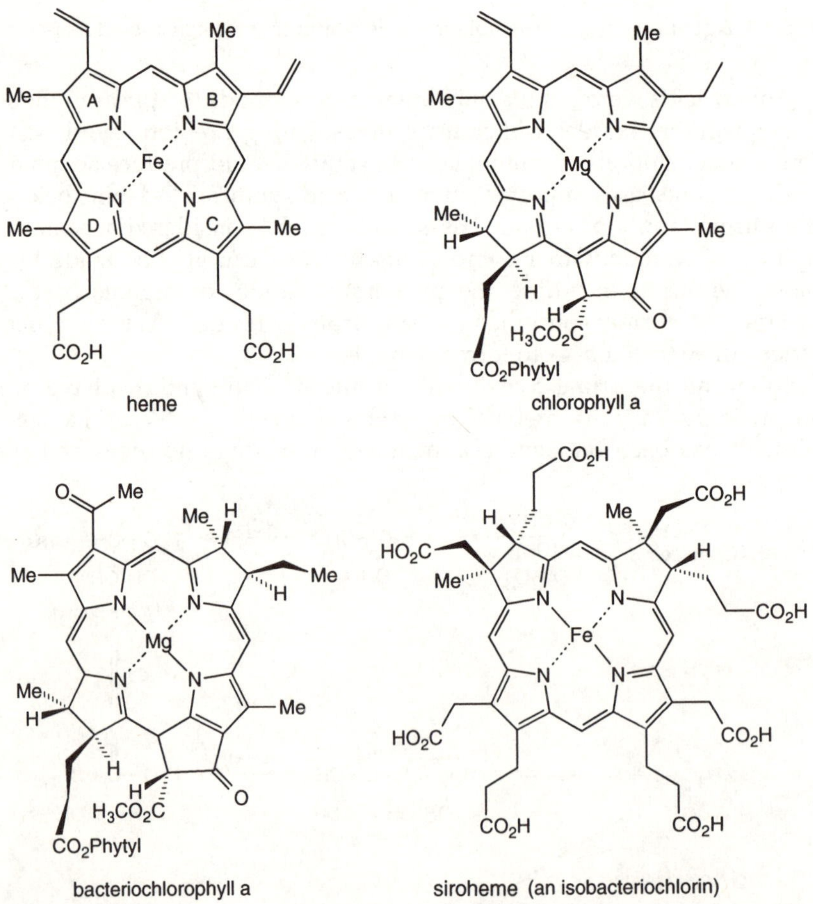

Fig. 2.5 The six main types of 'pigments of life' derived from uroporphyrinogen III.

2.2 Biosynthesis of the "pigments of life"

2.2.1 *From Glycine to Uro'gen III*

A careful look at the main "pigments of life" (as Professor Sir Alan Battersby from the University of Cambridge calls them), e.g. heme, chlorophyll, bacterio-chlorophyll, and vitamin B_{12}, shows that they have several features in common. They all consist of four pyrrole-type rings, with a greater or lesser degree of unsaturation (depending on the macrocycle), and they have a rather unusual pattern of substituents on their peripheral β-carbons.

Protoporphyrin IX, for example, has three kinds of β-substituent; methyl groups, vinyl groups, and propionic acid side chains. Observe the position of the methyl groups as we move clockwise around the macrocyle. For three of the pyrrole subunits (A, B, and C), the methyl group appears on the same β-carbon, but on the fourth (D) pyrrole, the methyl group is on the opposite β-carbon (see Figure 2.5). It is as if in making the macrocycle, nature strings four pyrrole units together nose to tail, and then perversely, before

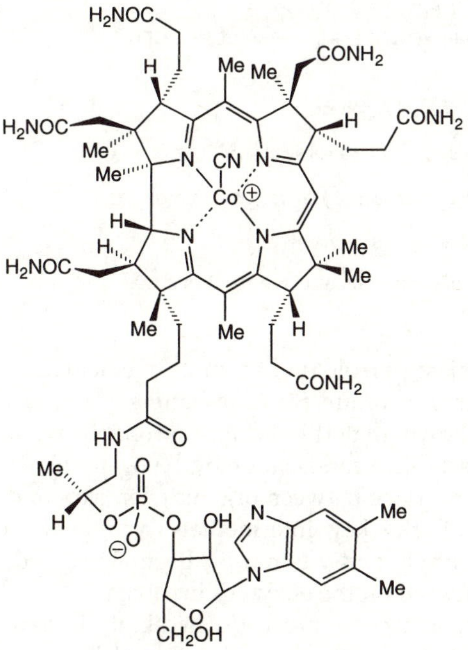

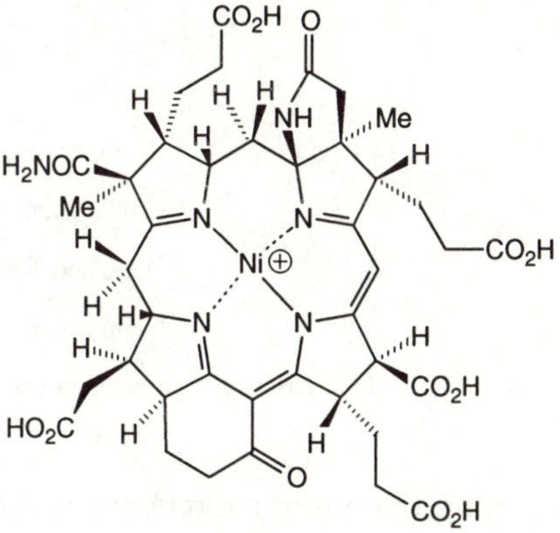

coenzyme F430

cyanocobalamin or vitamin B 12

Phytyl =

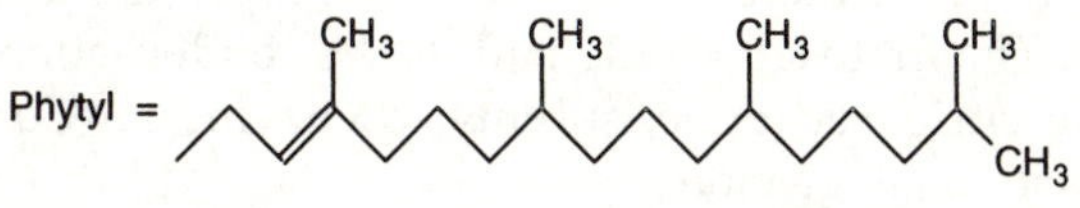

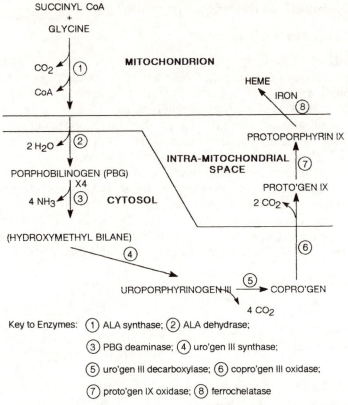

SUCCINYL CoA
+
GLYCINE

CO_2 ① **MITOCHONDRION**

HEME

CoA

IRON
⑧

2 H_2O ②

PROTOPORPHYRIN IX

INTRA-MITOCHONDRIAL SPACE ⑦

PORPHOBILINOGEN (PBG)
X4

PROTO'GEN IX

4 NH_3 ③ **CYTOSOL**

2 CO_2

(HYDROXYMETHYL BILANE)

④

⑥

UROPORPHYRINOGEN III ⑤ ► COPRO'GEN

4 CO_2

Key to Enzymes: ① ALA synthase; ② ALA dehydrase;

③ PBG deaminase; ④ uro'gen III synthase;

⑤ uro'gen III decarboxylase; ⑥ copro'gen III oxidase;

⑦ proto'gen IX oxidase; ⑧ ferrochelatase

Fig. 2.6 Overview of heme biosynthesis and where it occurs.

closing them into a circle, turns the last pyrrole unit around. In elucidating the biosynthetic pathway to the pigments of life, Sir Alan Battersby's group at Cambridge, and others, have shown that this is most probably what happens.[7] The work has taken nearly 40 years, Battersby being involved since 1968. The research involves a marriage between organic chemistry and biochemistry. For not only have all the key intermediates on the bio-synthetic pathways to heme, chlorophyll, and vitamin B_{12} been elucidated, much information has also been garnered on the enzymes involved.

In mammalian tissues, heme biosynthesis proceeds in eight discrete enzyme-catalysed steps (Figure 2.6). These steps are distributed between two parts of the cell, the mitochondrion, where the major metabolic reactions take place, and the cytosol, which is simply the aqueous phase of the cytoplasm. The first step on the path is the condensation of an inter-mediate from the citric acid cycle, succinyl coenzyme A (CoA), with the simplest amino acid of all, glycine.

(NOTE. The citric acid cycle is also known as the Krebs cycle, after its discoverer Hans Krebs. It is a cyclical series of biochemical reactions which is fundamental to the metabolism of aerobic organisms. The enzymes for the Krebs cycle are located in the mitochondria and are in close association with the enzymes involved in the electron-transport chain, making the Krebs cycle a kind of biochemical switching station in the complex system of cellular metabolic pathways. It links together degradation, energy production, and biosynthesis.) The fact that glycine was the start of heme biosynthesis in animals was discovered by David Shemin who, in 1945, swallowed 66 g of ^{15}N-labelled glycine over a three-day period. He took regular blood samples from himself, isolated the heme, and found that the ^{15}N atoms ended up in the porphyrin.[8] Shemin also found that the radiolabel was rapidly incorporated and then remained at a constant level for about 120 days (the approximate lifetime for a red blood cell), after which time it started to decline.

The first product formed by condensation of glycine and succinyl CoA is *5-aminolaevulinic acid* or ALA. This reaction is catalysed by the enzyme *ALA synthase* and requires the cofactor, pyridoxyl phosphate. ALA synthesis occurs entirely within the matrix of the mitochondrion and is very tightly regulated by control of the enzyme that makes it, control of the transport of the enzyme from the cytosol into the mitochondrion, and finally by inhibition of ALA synthase by the final product, heme.

The route to ALA is called, naturally enough, the Shemin route. ALA is a much better precursor of heme than glycine. This was shown by making ALA labelled with both ^{15}N and radioactive ^{14}C. Incubating this labelled ALA with the enzymes responsible for heme metabolism showed that all the ^{15}N and ^{14}C ended up in the heme.

Recently, however, it has been discovered that plants, algae, and anaerobic bacteria make ALA another way; directly from *glutamic acid*[9] via *2-oxoglutarate*, in what is called the C_5 route. It is possible that this C_5 route is an ancestral form of ALA biosynthesis, suited to organisms that did not have complete citric acid cycles and thrived in the early anaerobic atmosphere. Only later on, when oxygen levels had risen and the aerobes were dominant, did the Shemin route to ALA appear. Chloroplasts are thought to have evolved from organisms that had the same origin as the ancient anaerobic cyanobacteria. In order to survive in an oxygenated atmosphere, they went into symbiotic partnership with aerobic organisms, ultimately becoming incorporated within plant cells. This would explain why plants still have the C_5 pathway to ALA.

The next step in heme biosynthesis, is the condensation of two molecules of ALA, via the enzyme *ALA dehydratase*, to give the pyrrole called *porphobilinogen* or PBG. This compound was first identified in the urine of patients suffering from one of a family of diseases of porphyrin metabolism, called porphyrias. Labelled PBG has been efficiently incorporated into heme using chicken blood as the biosynthetic model system. What is interesting about the biosynthesis of PBG is that it is very similar

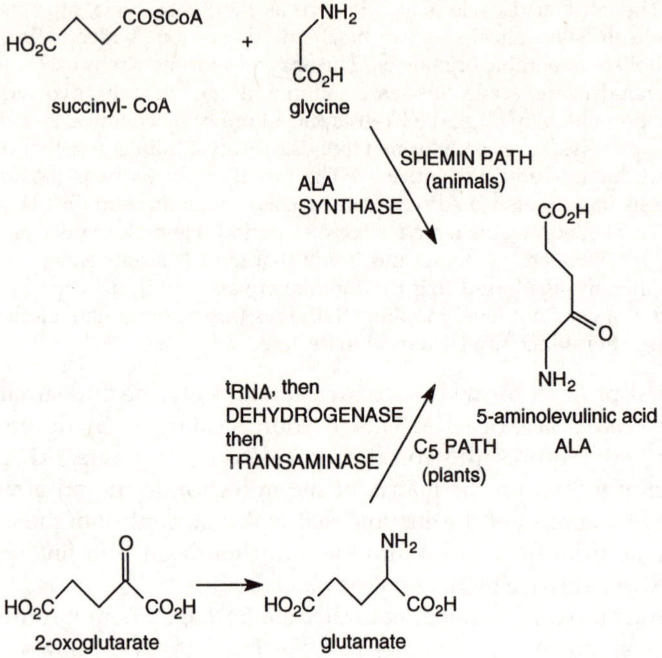

Fig. 2.7 The two paths to ALA.

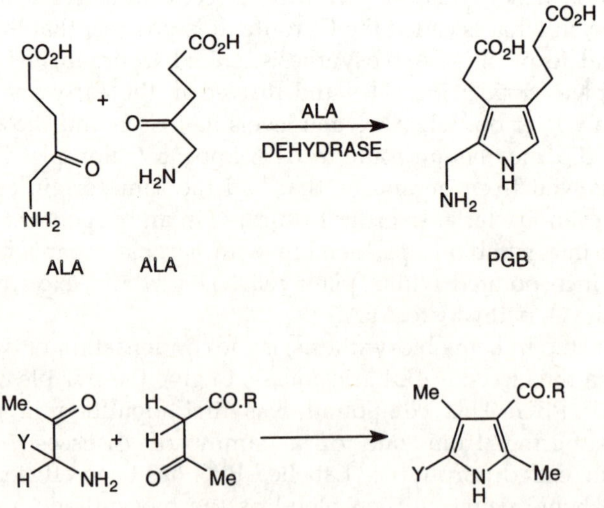

Fig. 2.8 Formation of PBG and a typical Knorr pyrrole synthesis.

to one of the main routes used to make pyrroles in the laboratory, called the Knorr pyrrole synthesis.

PBG formation is followed by condensation of four PBG molecules to give *uroporphyrinogen III* (porphyrinogens are reduced porphyrins; the fully oxidised macrocycle only appearing at the end of the biosynthetic pathway: meanwhile, the name being too long to repeat all the time, is abbreviated to *uro'gen III*). It is here that the unexpected β-substitution pattern is first encountered. The placement of the acetic acid side chains mirrors the positioning of the methyl groups in protoporphyrin IX. The condensation of the four PBG units also requires not one, but *two* enzymes; *PBG deaminase* and *uro'gen III cosynthetase*.

What has fascinated those working in this field about this all important step, is that the four identical PBG units are joined together in a head-to-tail

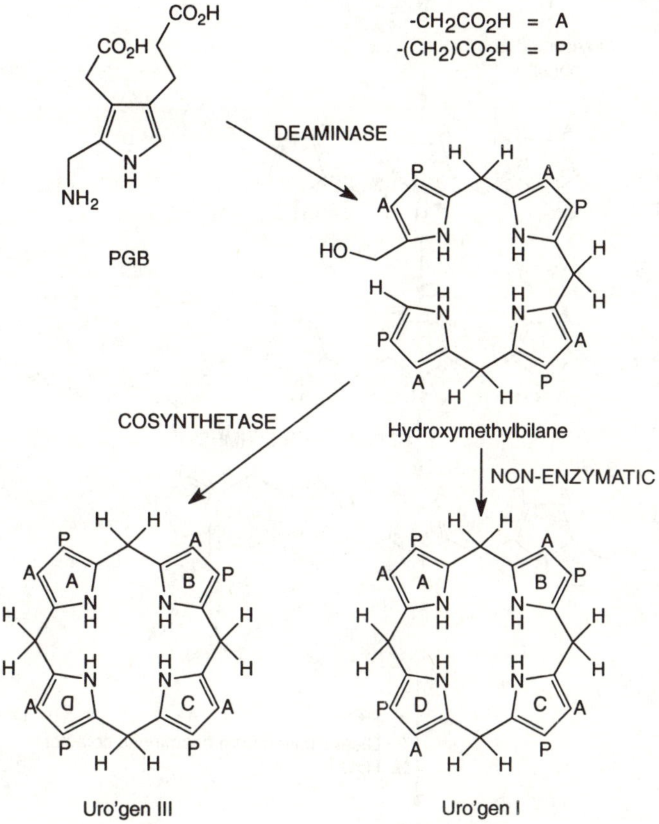

Fig. 2.9 & 2.10 Enzymatic formation of uro-gen III. Note intramolecular rearrangement of ring D.
Non-enzymatic formation of uro'gen I.

fashion (rings A to D), but then one of them (ring D) appears to have been turned around to give the type III isomer in 100% yield. This requires both enzymes because in the absence of uro'gen III cosynthetase, only uro'gen I is produced and this cannot be converted by cosynthetase into the type III isomer.

To give some idea of just how much this step in porphyrin biosynthesis has taxed the ingenuity of chemists, at lest 25 hypothetical schemes have been produced to explain it. The main picture to emerge, ultimately from

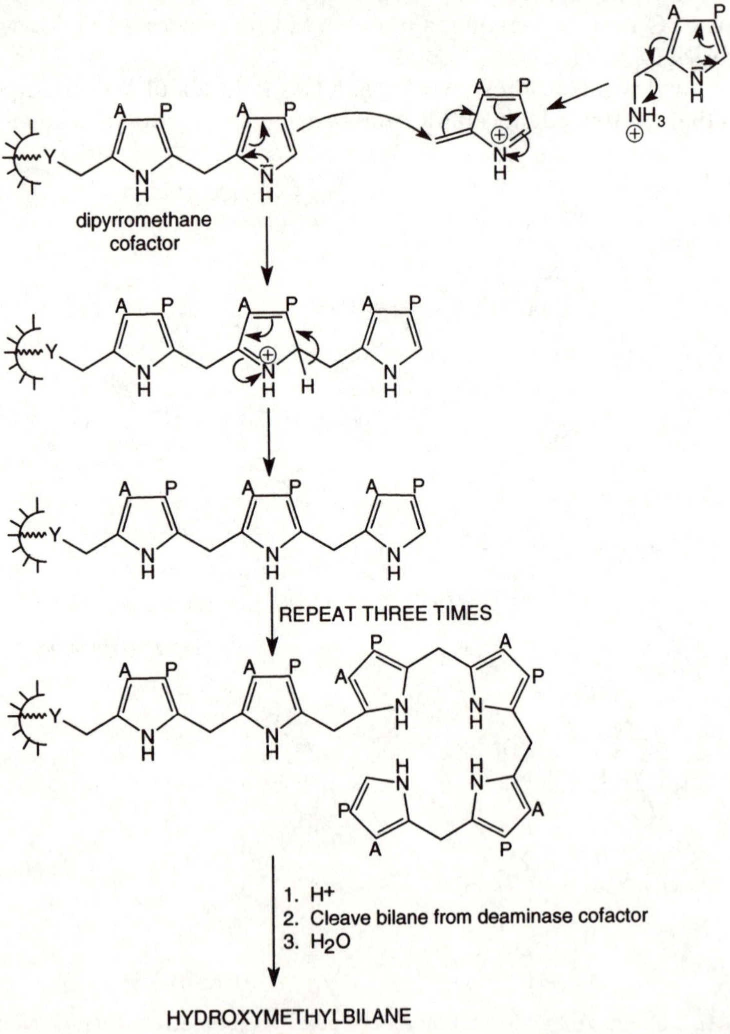

Fig. 2.11 Formation of hydroxymethylbilane via a dipyrrylmethane cofactor on deaminase.

Prof. Battersby's laboratory in Cambridge, is that deaminase has a *dipyrryl-methane cofactor* which consists of two previously formed PBG units. This cofactor acts as an anchor on to which four more PBG units are attached, to form an open chain *hydroxybilane* intermediate. After cleavage from deaminase, this intermediate acts as the substrate for cosynthestase.

Battersby was able to show that cosynthetase literally turns ring D around. As this is a story that makes a fine piece of chemical detective work, we shall spend a little time on it. Battersby and his team made PBG labelled in two positions with the NMR-active carbon atom, ^{13}C, which, like ^{1}H, ^{19}F, and ^{31}P, has a nuclear spin of ½. When this double-labelled PBG is incubated with deaminase on its own, uro'gen I is produced which is isolated as uroporphyrin I octamethylester, labelled with ^{13}C. NMR spectroscopy detects the carbon-13 signals from the *meso*-carbons which show strong spin–spin coupling from an adjacent ^{13}C from a different incorporated PBG molecule, and a smaller coupling from a more distant ^{13}C from the same incorporated PBG.

By diluting the double-labelled PBG with four parts of unlabelled PBG, and then incubating the mixture with deaminase and cosynthetase (and other enzymes that take the labelled uro'gen III through to protoporphyrin IX labelled in exactly the same carbon atoms), most of the uro'gen III

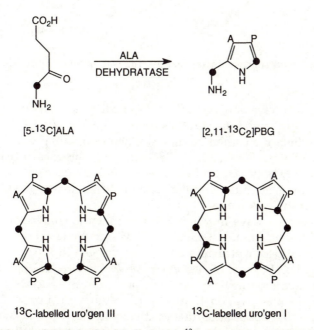

Fig. 2.12 Synthesis of PBG double labelled with ^{13}C. Use of this labelled PBG shows how the four PGB units are connected in uro'gen III and uro'gen I.

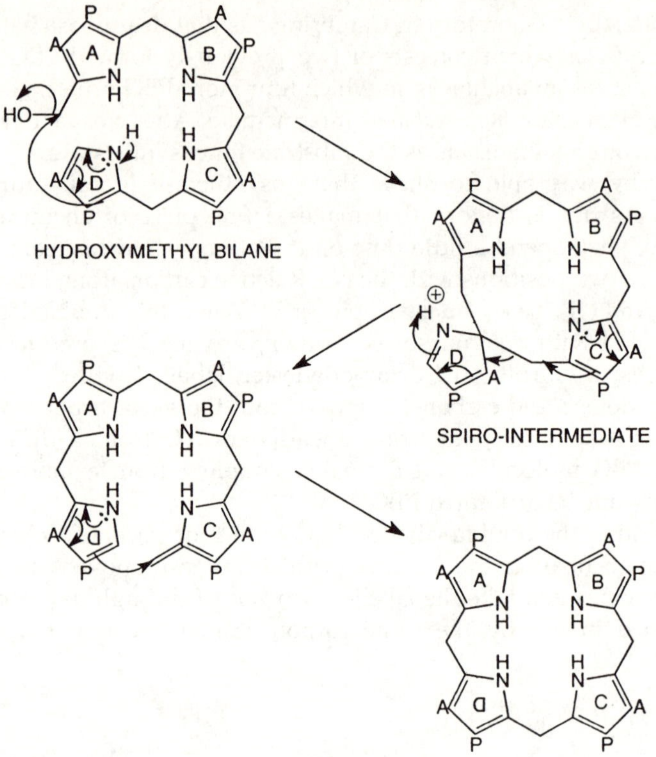

Fig. 2.13 Action of cosynthetase causes ring D to flip via a spiro-intermediate.

molecules will have only one pair of labelled ^{13}C atoms. Consequently, the ^{13}C signals for three of the *meso*-positions (C-5, C-10, and C-20) show only a weak coupling, the carbon atom, C-15, shows a strong coupling from an adjacent ^{13}C atom *from the same PBG molecule*. This means that the whole pyrrole ring D is turned around, it is the only pyrrole ring to rotate, and that it occurs as an intramolecular reaction, as no exchange with unlabelled PBG precursors took place.

There are two ways in which the cosynthetase enzyme could turn the fourth PBG unit around. Battersby believes an enzyme-bound spiro-intermediate is formed, which goes on to form uro'gen III.[10] An alternative proposal, by Ian Scott at Texas A & M University, is that the cleaved hydroxybilane intermediate self-assembles to form a *lactone* via the acetic acid side chain at position 17 in ring D.[11] Although support for this view comes from the observation that synthetic bilanes without this side chain will not undergo enzyme-catalysed rearrangement to the type III isomer, the weight of experimental evidence seems to support Battersby's spiro-

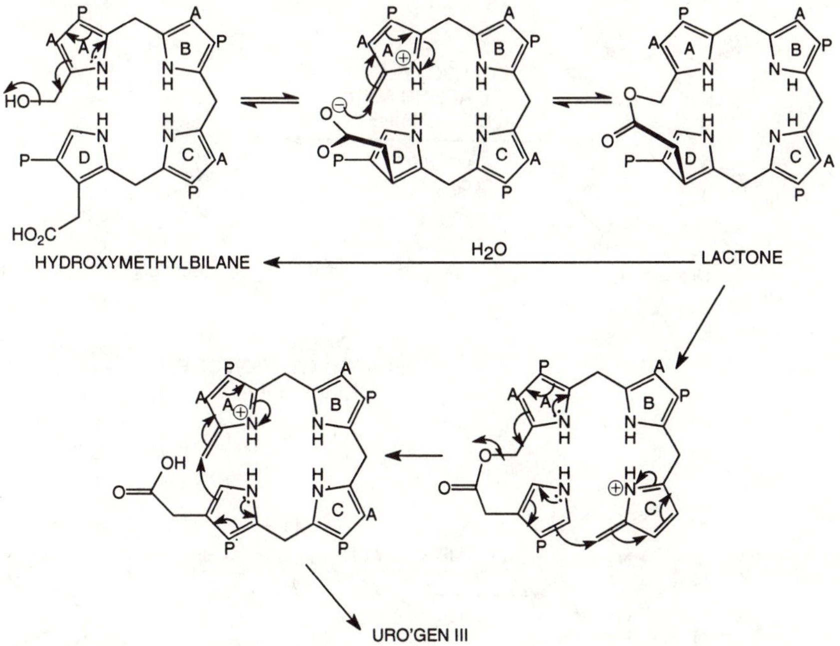

Fig. 2.14 Scott's alternative mechanism for the flipping of ring D via lactone formation between the acetic acid group of ring D and ring A.

intermediate. Whatever the mechanism for this rearrangement, the product, uro'gen III, is the starting point for the biosynthesis of all the pigments of life.

2.2.2 *Uro'gen III to heme*

The four acetic acid side chains are decarboxylated (by the enzyme *uro'gen decarboxylase*) to methyl groups to give *copro'gen III*, followed by the oxidative decarboxylation of the two propionic acid side chains on rings A and B to vinyl groups (via the enzyme *copro'gen oxidase*) giving *proto'gen IX*. Oxidation of proto'gen IX to *protoporphyrin IX* is then accomplished by the enzyme *proto'gen oxidase*, followed by iron insertion catalysed by a very similar enzyme called *ferrochelatase*.[12] Both enzymes are bound to the inner mitochondrial membrane.

The heme is then inserted into a protein to become a hemoprotein. In the case of non-covalently bound hemes (i.e. in hemoglobin and myoglobin, cytochromes *b* and P450, catalase, and peroxidases), the association between heme and apoprotein is spontaneous and does not require a catalyst. In contrast, the covalently bound hemes (e.g. cytochrome *c*) require

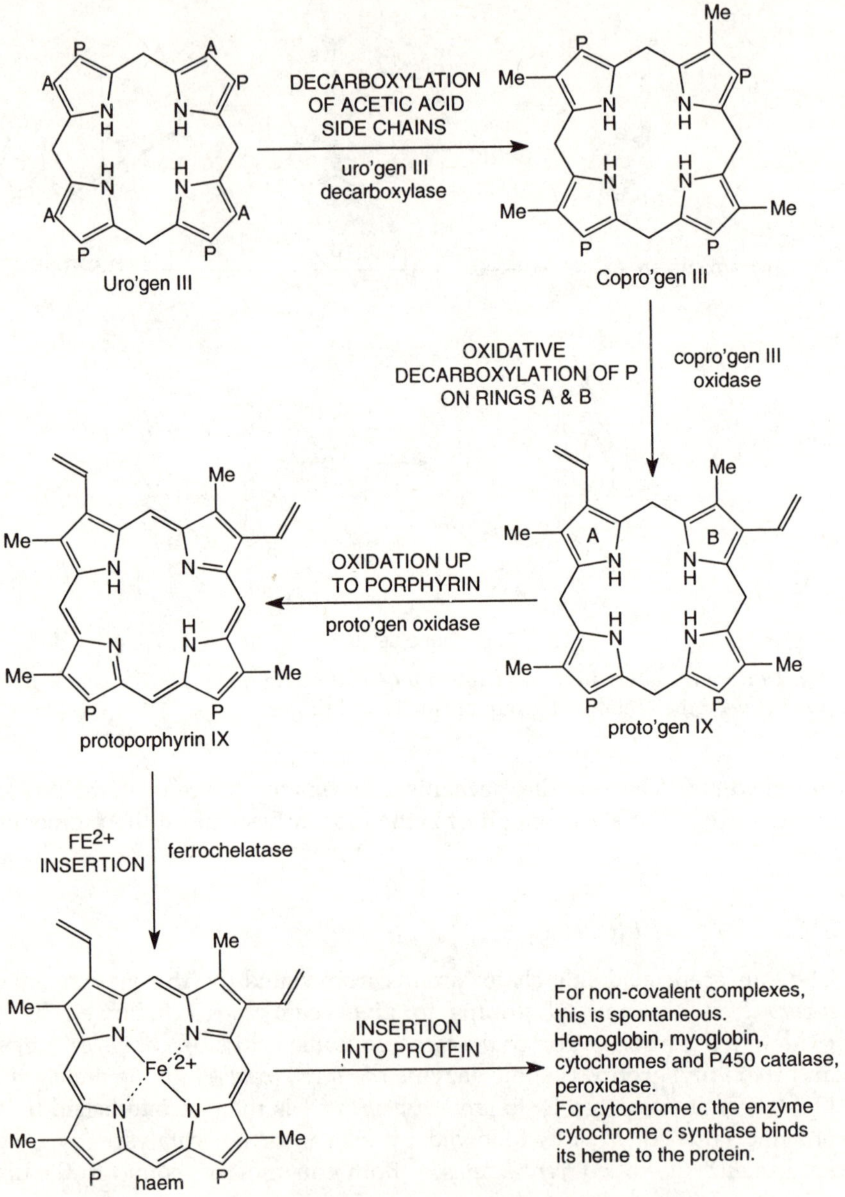

Fig. 2.15 From uro'gen III to heme.

enzymic catalysis to form thioether links to the heme vinyl groups. The enzyme that performs this operation, cytochrome-*c* synthase, is located either on the outer face of the inner mitochondrial membrane, or in the intermembrane space.[13] This enzyme might also be implicated in bringing the cytochrome *c* apoprotein across the outer mitochondrial membrane from the protoplasm.[14] What is thought to happen is that the apoprotein spontaneously inserts into the membrane, binding to a protein on the inner surface. When the heme attaches to make cytochrome *c*, a conformational change occurs which forces the cytochrome inside, where it migrates to its final location on the outside of the membrane.

2.2.3 Uro'gen III to chlorophyll

The biosynthesis of chlorophylls[15] diverges from that of heme at the metal insertion stage. Magnesium is inserted into protoporphyrin IX by a little-known enzyme, called *magnesium chelatase*, followed by esterification of the propionic acid side chain on C-13 by transfer of a methyl group from

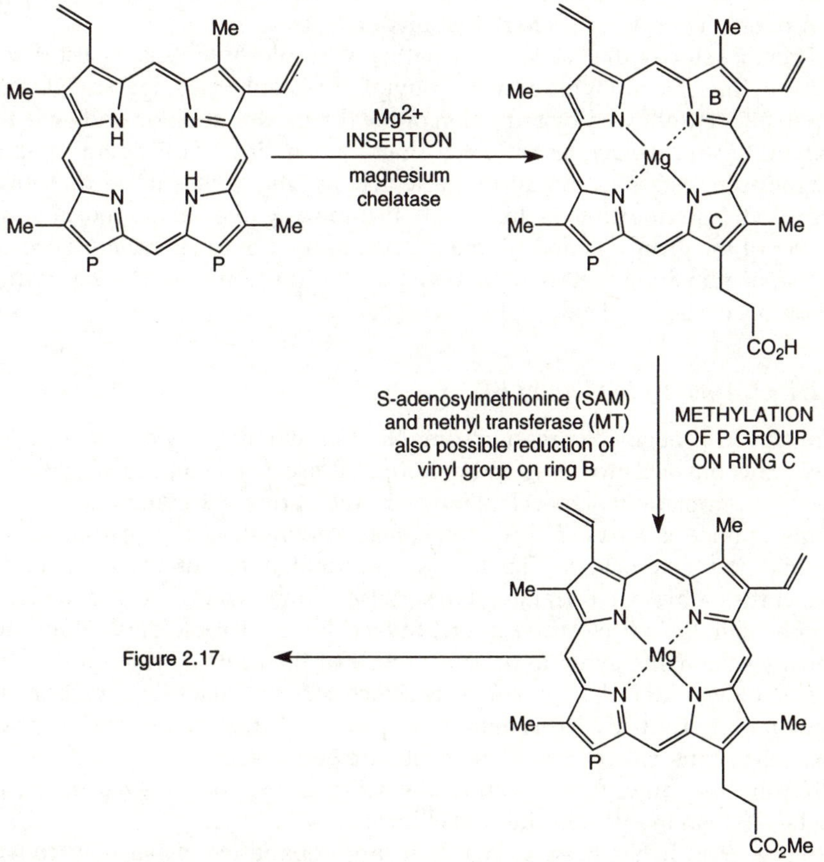

Fig. 2.16 Chlorophyll biosynthesis from protoporphyrin IX to magnesium protoporphyrin IX-13-methylpropionate.

S-adenosyl—methionine (SAM). At this point, the vinyl group at C-8 may or may not be reduced to an ethyl group, depending on the species of plant and whether it is day or night. Alternatively, this reduction can happen later on at the chlorphyllide stage.

The next step is the conversion of the C-13 propionic acid side chain into a β-keto ester and its cyclisation to the characteristic exocyclic ring of the chlorophylls, to give *protochlorophyllide*. This step is catalysed by an *oxidative cyclase* and requires oxygen and NADPH. Interestingly, the specificity of this enzyme is such that the vinyl group at C-3 must be present for this cyclisation to occur. Also, magnesium or zinc porphyrins are good substrates for this enzyme, but nickel, copper, or the metal-free porphyrins are not. Protochlorophyllides then undergo photochemical reduction, via the enzyme *protochlorophyllide reductase* which is NADPH dependent, to give *chlorophyllide*. There is also evidence that in some plants, at least, there are non-photochemical routes to reduction of ring D.

The final step in the biosynthetic pathway to chlorophylls is the esterification of the chlorophyllides with phytyl diphosphate or geranylgeranyl diphosphate (followed by reduction of the three extra double bonds) via the enzyme *chlorophyll synthetase*. Once again, when zinc or cadmium replaces magnesium, esterification is unaffected. However, when nickel or copper are used, esterification is hindered. Presumably the labile coordinating power of the group IIa and IIb metals, as opposed to the more inert coordination of transition metals, is involved in the function of the two enzymes, oxidative cyclase and chlorophyll synthetase.

2.2.4 Uro'gen III to vitamin B_{12}

Uro'gen III probably represents some kind of evolutionary crossroads. The pathways onwards to heme and chlorophyll involve a controlled oxidation up to an aromatic macrocycle. However, the pathway to vitamin B_{12} coenzyme is non-oxidative and involves methylation at various carbon atoms around the macrocycle. The fact that oxidation is not involved in B_{12} biosynthesis almost certainly dates it to a time when oxygen had not appeared in the earth's atmosphere. In fact, some microbiologists think that corrins—the name given to the macrocycle at the heart of vitamin B_{12}—are functionally older than porphyrins, since strictly anaerobic bacteria are known that produce corrins but not porphyrins.[16] In addition, the biological task of corrins is catalysis of biosynthetic steps, such as the synthesis of DNA and the amino acid, methionine. Corrins are not involved in energy metabolism as are porphyrins and chlorins.

Since 1926, it has been known that large quantities of partially cooked liver can help cure pernicious anaemia. This is an often fatal, complex disease, in which there is a deficiency in red cells and hemoglobin formation,

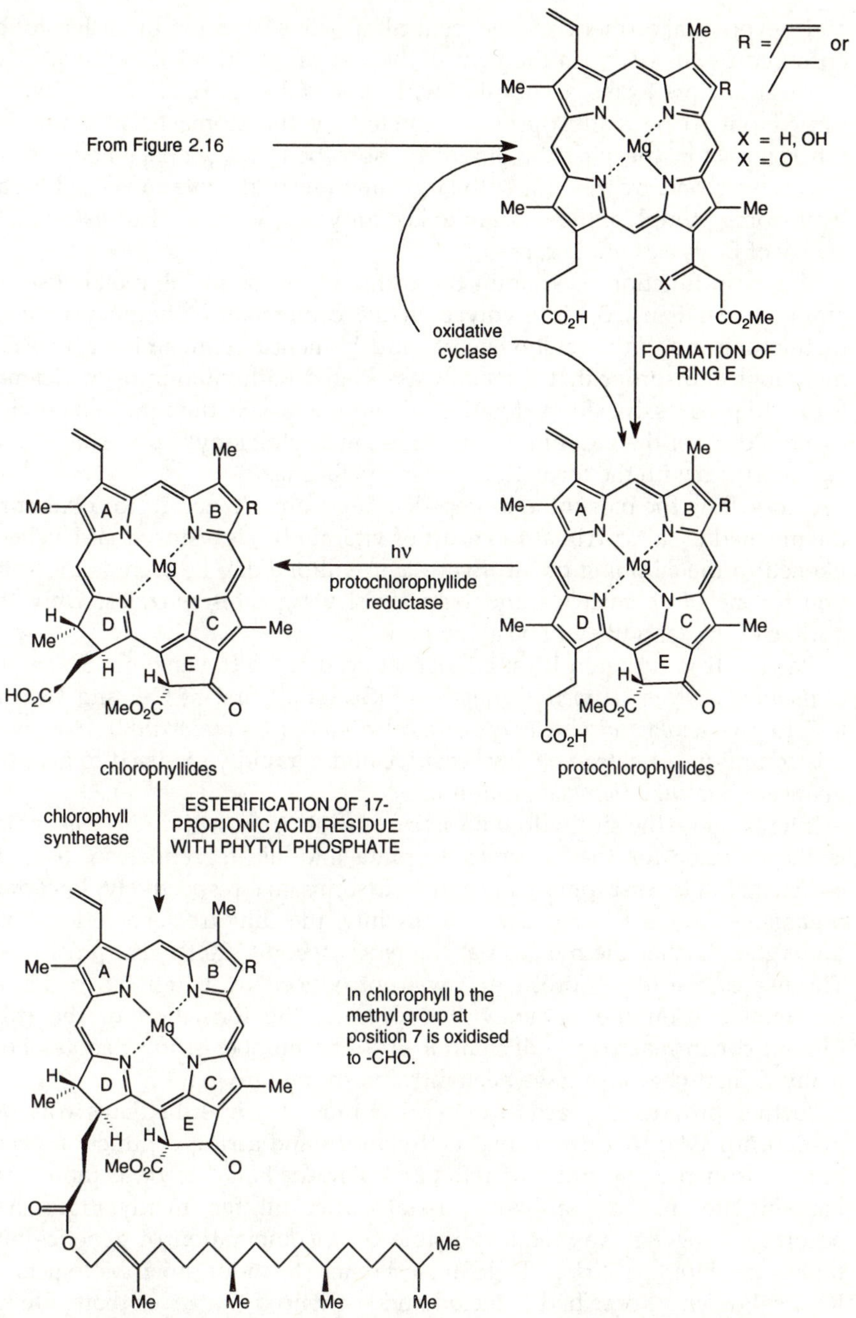

Fig. 2.17 Chlorophyll biosynthesis: ring E formation, reduction of ring D, and phytyl esterification.

and severe impairment of the central nervous system. In order to be absorbed by the body (in the ileum), the vitamin B_{12} (initially known as an extrinsic factor) forms a complex with a so-called intrinsic factor, which turned out to be a glycoprotein secreted by the stomach. Without this intrinsic factor, the vitamin cannot be absorbed and a deficiency results. This is overcome by treating with large amounts of the vitamin. The human liver stores enough of the vitamin to last for years, so that a true *dietary* deficiency of B_{12} is extremely rare.

The main function of vitamin B_{12} is thought to be in the metabolism of amino acids. Thus, B_{12} is involved in the conversion of homocysteine to methionine and in the catabolism of some branched-chain amino acids. The neurological disorder that is usually associated with vitamin B_{12} deficiency is due to progressive demyelination of nervous tissue, thought to be owing to a build up of the vitamin B_{12} substrate, methylmalonyl CoA. This probably interferes with the formation of the myelin sheath.

Unravelling the biosynthetic steps leading from uro'gen III to cobalamin, the finished cobalt corrin at the heart of vitamin B_{12}, is a story that has been likened to the climbing of Mt. Everest. It will one day be seen as the solution to one of chemistry's and biochemistry's greatest puzzles. Only the outline of the "ascent" will be given here.

Starting from uro'gen III, methylation occurs, via the enzyme *S-adenosyl methionine uro'gen III methyl transferase* (SUMT),[11] first at C-2 and then at C-7, to give a *dihydroisobacteriochlorin* (also known as *precorrin 2*). However, this compound is extremely air sensitive and is rapidly oxidised to give the *isobacteriochlorin* (or *sirohydrochlorin*).

Interestingly, the sirohydrochlorin can chelate iron to give *siroheme* which is the cofactor for the enzymes sulphite and nitrite reductase, used to reduce sulphite to sulphide and nitrite to ammonia, respectively, in certain organisms (e.g. *Esherichia coli*). Meanwhile, the dihydroisobacteriochlorin undergoes further methylation at the *meso*-carbon C-20 to give *precorrin-3*. The use of the term "precorrin" was introduced by Battersby to denote intermediates on the pathway that precede the formation of the fully formed corrin macrocycle of vitamin B_{12}. The number suffix denotes how many C-methyl groups have been introduced into uro'gen III.

Further progress towards discovering more B_{12} intermediates was not made until 1990. A further five methylations and a ring contraction occur but the sequence was unknown. In particular, the latter step was thought to happen late in the pathway, possibly after all the methylations had occurred. Progress was made possible by a combination of genetics and molecular biology. Prior to 1990, a French team of microbiologists at Rhone–Poulenc–Rorer had detected and sequenced the genes that code for the various enzymes that produce vitamin B_{12} in the organism, *Propionibacterium denitrificans*. Knowing the genetic codes for these enzymes, it was

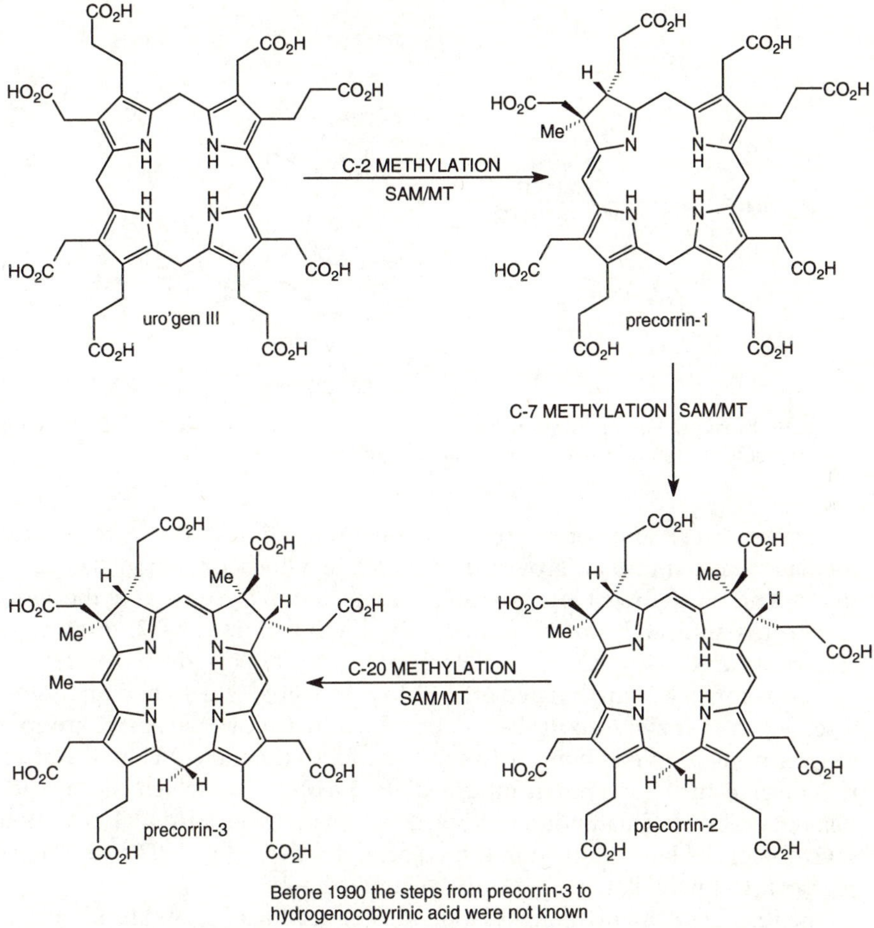

Fig. 2.18 Biosynthesis of vitamin B_{12}: uro'gen III to precorrin-3.

possible to produce genetically engineered strains of *P. denitrificans* in which certain of the B_{12} enzymes were overexpressed. This allows reasonably large quantities of B_{12} intermediates to build up, making their detection (usually by ^{13}C NMR spectroscopy, after feeding the bacteria with precursors labelled with ^{13}C) and structure determination a lot easier.

In 1990, a new intermediate was discovered. Its structure wasn't completely worked out until 1992, and it showed that ring contraction occurs well before all the methylations are completed. This intermediate was called *precorrin-6x* and since 1992 other intermediates before and after this one in the pathway have been discovered and their structures determined. We shall describe the pathway using the sequence of steps that is thought to occur at the moment.

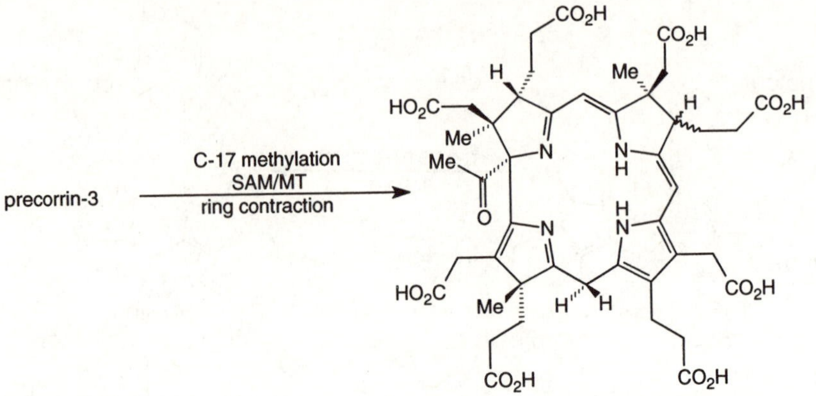

Fig. 2.19 Biosynthesis of vitamin B_{12}: precorrin-3 to precorrin-4, *c.* 1993, showing that ring contraction occurs earlier not later in the pathway.

The next steps after preocorrin-3 involve formation of a γ-lactone intermediate, *precorrin 3B* (discovered in 1994) in which oxidation has taken place. This is followed by extrusion of the methyl group and the *meso*-carbon (as an acetyl group), followed by methylation at C-17, to give the ring-contracted corrin-type macrocycle of *precorrin-4*, discovered in 1993.[17a] Next, precorrin-4 is methylated at C-11 to *precorrin-5*, which then undergoes a sequence of enzyme-controlled steps which (i) remove the acetyl group at C-1 (forming a double bond between C-1 and C-19) and (ii) C-methylate at C-1 to give the next known intermediate, *precorrin-6x*, which has already been mentioned. This undergoes hydride transfer from NADPH at C-19 to give one of the latest known intermediates, discovered in 1992 and dubbed *precorrin-6y* by the Battersby group at Cambridge.[17b]

The final steps involve methylation at *meso*-carbons C-5 and C-15, to give *precorrin-8x* followed by a [1,5]-sigmatropic shift of the methyl group from C-11 to C-12, decarboxylation of the acetic acid group at C-12, to give *hydrogenobyrinic acid*, and insertion of cobalt to give the completed metallo-corrin macrocycle, *cobyrinic acid*.

The point at which cobalt is inserted depends on the organism. Thus, *Propionibacterium denitrificans* inserts cobalt into the macrocycle at the hydrogenobyrinic acid stage, whereas in *P. shermanii*, cobalt insertion is thought to occur prior to ring contraction around the precorrin-3 stage.[17c] In a few years, a problem that was thought to have its ultimate solution some time in the 21st century, has been solved in the last decade of the 20th century.

Although all of these steps and their elucidation are extremely complex, there is growing evidence from attempts to synthesise the vitamin B_{12} structure chemically, that most of the structural elements in the molecule (even

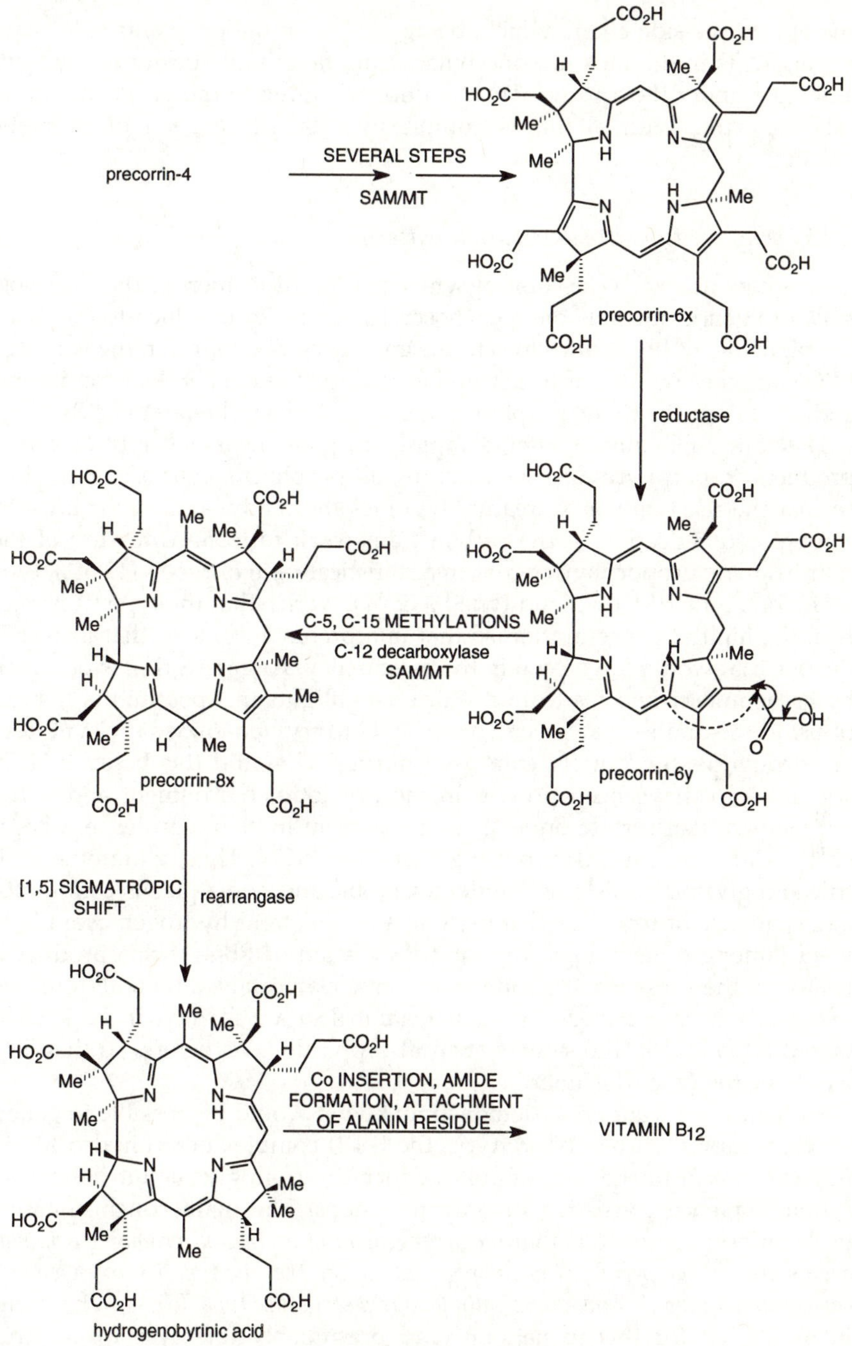

Fig. 2.20 Biosynthesis of vitamin B_{12}: precorrin-4 to vitamin B_{12}.

the nucleotide side chain, which, being placed on the propionic side chain of ring D, is in the most thermodynamically favourable position) will self-assemble under the appropriate conditions, leading to the proposition that this molecule, with all of its complexity, may have been of prebiotic origin.[16]

2.2.5 Why the type III substitution pattern?

This brings us to the question of why the type III isomer, of the four possible uro'gen isomers, is the one chosen by nature from which to construct the pigments of life. Some clues to the answer can be found in the way that PBG combines with itself under non-enzymatic (and anaerobic) conditions, and what happens to the porphyrinogens under the influence of acid.

The four PBG units combine rapidly to give the kinetically favoured product, uroporphyrinogen I. However, all porphyrinogens are acid labile so that the reaction course readily becomes thermodynamically controlled through isomerisation of the carbon framework to form a mixture of the four isomeric uroporphyrinogens in a statistical ratio of type I (12.5%), type II (12.5%), type III (50%), and type IV (25%). Notice that the type III isomer is in the highest concentration so that in a prebiotic context, that isomer is the one that would have been in the ascendency. An interesting experiment by Eschenmoser and his team at Zurich highlights this possibility. Instead of using the amino acid precursors of PBG to try non-enzymatic syntheses of porphyrins, the Zurich team used nitriles. They did this because HCN was likely to have been present in the prebiotic environment and it has been shown that certain organic species present in biomolecules, e.g. bases such as adenine, are easily put together from HCN. Thus, glutamine dinitrile and glycine nitrile react, under anaerobic and non-aqueous conditions, in a sequence of reactions that extrude ammonia and hydrogen cyanide to yield (among other things) the dinitrile version of PBG. If this product is heated in the presence of a montmorillonite clay catalyst, then all four isomeric uroporphyrinogens (as their octanitriles) are formed in high yield (about 80%) and in the same statistical proportions as before, so that 50% would be the type III isomer.[6]

Eschenmoser went on to demonstrate how it would be possible to generate the simplest chromophore type, the Fe(II) complex of sirohydrochlorin (the cofactor of nitrite and sulphite reductase), from glycine nitrile in a reaction sequence consisting of only five steps. The main conclusion that Eschenmoser comes to is that *the arrangement of acetic and propionic acid side chains around the periphery of the pigments of life (i.e. the type III arrangement) corresponds to the thermodynamically favoured structure type*. It has to be borne in mind that for this to happen (and presumably for there to be large amounts of porphyrins and other tetrapyrroles available as prebiotic

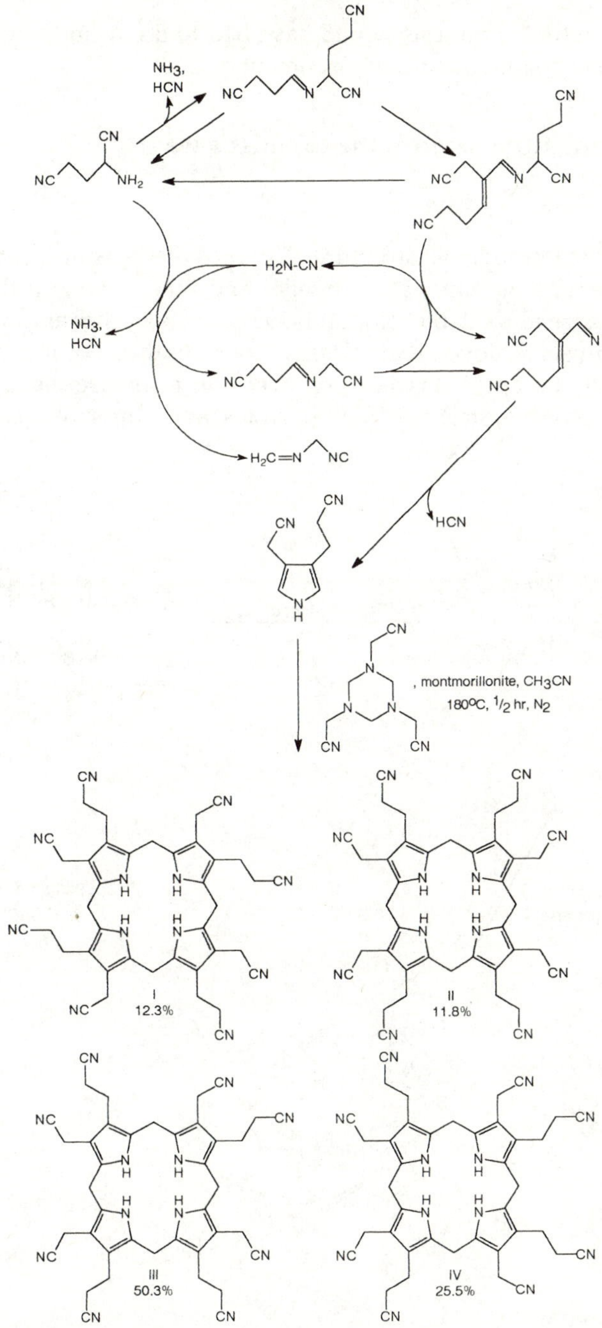

Fig. 2.21 Albert Eschenmoser's potentially prebiotic route to uroporphyrinogens (octanitriles): note preponderance of type III isomer.

materials), then their synthesis would have had to occur not only anaerobically, but also in a non-aqueous environment.

2.3 Porphyrins from scratch: the chemist's way

2.3.1 Introduction

The macrocyclic tetrapyrrole structure of porphyrins was first suggested in 1912 by Küster. At the time, Hans Fischer and others thought that such a ring was too large to be stable. Not until 1929, when Fischer and his Munich School of chemists achieved their classical total synthesis of heme, was this structure finally accepted. Fischer then went on to devise many synthetic routes to porphyrins starting from pyrroles and dipyrrolic compounds

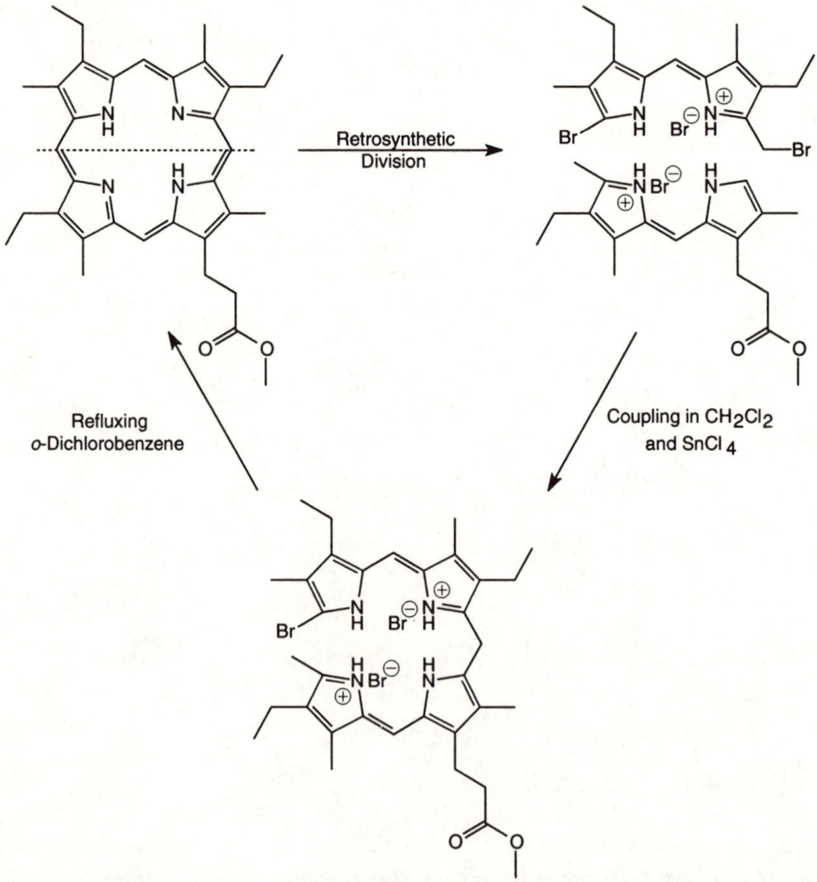

Fig. 2.22 Retrosynthetic division of etio 1 to protected pyrroles.

called dipyrromethenes. Subsequently, others, such as George Kenner, Tony Jackson, and Kevin Smith at Liverpool University and Alan Johnson at Nottingham in the UK and R.B. Woodward in the USA, devised new synthetic procedures involving dipyrrylmethanes, a,c-biladienes, bilenes, and oxobilanes. As the literature on these syntheses is already immense,[18] no attempt will be made to study these routes in detail. Instead, suitable examples will be given that will hopefully demonstrate the art of these great chemists.

Porphyrin chemistry from scratch is not a task to be undertaken lightly. I found that out to my cost as an eager young postgraduate student with the late Professor George Kenner. My project was on hemoprotein model compounds so, to get me started, a fairly "simple" porphyrin was chosen for me to synthesise via Alan Johnson's a,c-biladiene route.[19] The porphyrin, a propionic acid derivative of etioporphyrin type I (or, to give it is systematic name *8,13,18-triethyl-2,7,12,17-tetramethylporphyrin-3-propionic acid*, and its semi-systematic name *13-deethyletioporphyrin I-13-propionic acid*) was to be made from two dipyrromethene halves. Continuing this retrosynthetic analysis, these were, in turn, to be constructed from protected monopyrrolic starting materials.

The two dipyrromethenes were to be linked together in a Friedel–Crafts reaction (using stannic chloride) to form the a,c-biladiene. The final step to the porphyrin required refluxing this biladiene in 1,2,-dichlorobenzene at 180 °C. This straightforward preparation, which should have taken three to four weeks (bearing in mind my complete inexperience in porphyrin chemistry at the time) and yielded a good couple of grams of the required porphyrin, actually took six months; by which time only 100 mg of porphyrin had been amassed, and I was considering an alternative career to chemistry.

2.3.2 Porphyrins from heme and chlorophyll

Before we embark on some of the more elegant porphyrin syntheses, it should be pointed out that nature provides a huge store of porphyrins from which others may be derived. Thus, heme from blood and chlorophyll from plants can both be used to generate a wide variety of porphyrins. Starting from heme, Figure 2.24. shows what is possible.[20]

Similar diagrams can be drawn to show how chlorophyll can be used as a starting material for a wide range of porphyrins and chlorins, with and without *meso*-substituents.

Of course, it is necessary to acquire large quantities of hemin (i.e. Fe(III) protoporphyrin IX chloride) and chlorophyll. The first will involve a trip to the local abattoir for several gallons of fresh blood, while the second will involve an excursion to the local vegetable market for several kilos of spinach.

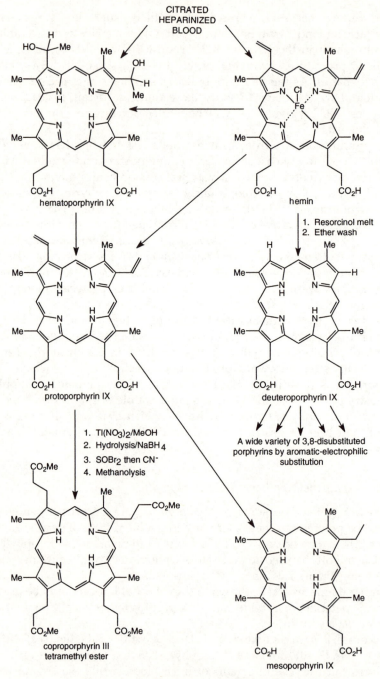

Fig. 2.23 Porphyrins derived from blood.

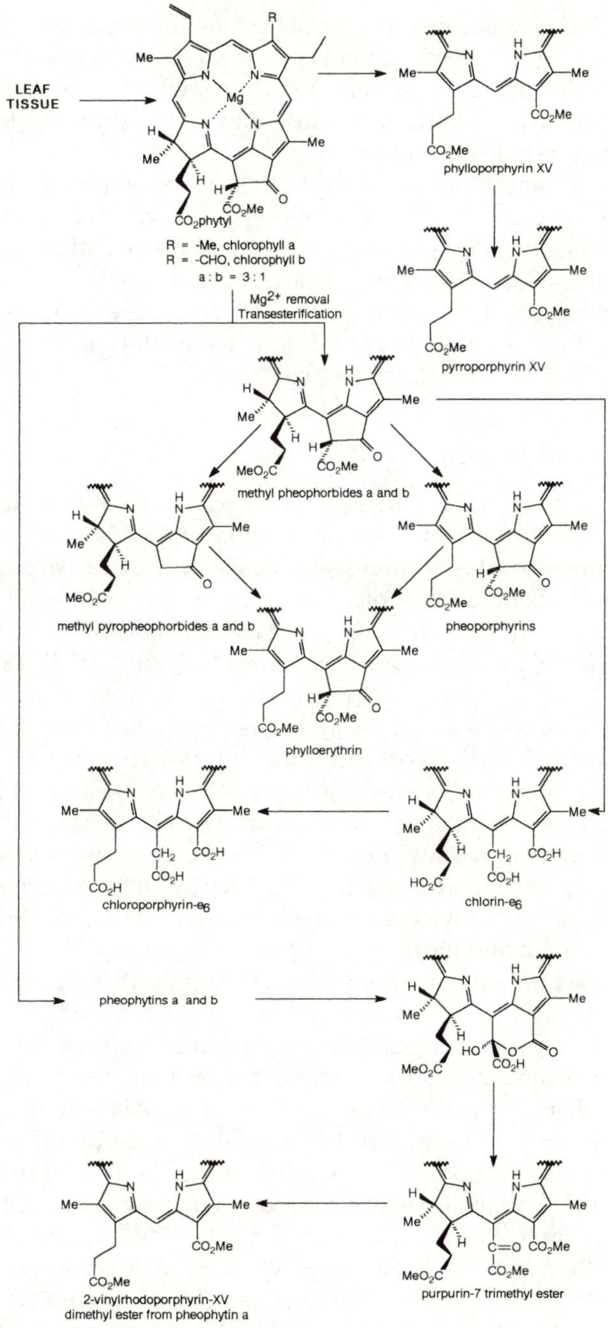

Fig. 2.24 Chlorins and porphyrins derived from chlorophyll (trivial names used).

After straining, heparinization, or defibrination, hemin is separated from blood by pouring the blood into hot acetic acid containing sodium chloride. On cooling, the hemin separates. Another method is to precipitate the globin protein with strontium chloride and concentrate the filtrate, from which hemin crystallises.

Chlorophyll *a* and *b* can be extracted from leaves simply by boiling them in methanol or acetone. The two chlorophylls are separated by column chromatography on silica gel, along with other photosynthetic pigments.

Of course, the only kinds of porphyrins that can be synthesised from heme and chlorophyll will have the type IX configuration of β-substituents. In order to obtain porphyrins with different substitution patterns, it is necessary to build porphyrins from scratch.

2.3.3 Porphyrins from pyrroles

Ultimately, *all* porphyrin syntheses start from pyrroles (which themselves need to be made), so that a special section with this title might appear pointless. However, this section really deals with those porphyrins where the number of chemical manipulations that the pyrrole has to go through prior to porphyrin formation, is minimal: the porphyrin is synthesised directly from the final pyrrole. This means that such synthetic routes are usually confined to those porphyrins with a symmetrical arrangement of peripheral substituents, e.g., 5,10,15,20-*meso*-tetrakis(aryl)porphyrin (e.g. TPP) and 2,3,7,8,12,13,17,18-octa-substituted porphyrins (e.g. OEP).

The narrow range of this synthetic route belies its practical utility. Strange as it may seem, most chemists are not porphyrin chemists, so if their chemical interests happen to stray into one of the many areas where porphyrins play a part (more of that in Chapter 7), they want to be able to use porphyrins that are relatively cheap and easy to make. TPP and OEP are probably the best known and most used porphyrins in the world, so much so, that many manufacturers of fine chemicals will supply them. However, until a chemist has separated a gram of porphyrin by column chromatography and seen the purple lustrous crystals made with his or her own hands (and spilt all over their bench and lab coat, or coated over the inside of their rotary evaporator), then they really cannot say they've worked with porphyrins.

TPP is the simplest porphyrin to make. It is made by refluxing pyrrole with benzaldehyde in propionic acid for half an hour. The purple solid is filtered off after cooling. It is washed (to free the solid from black, soluble poly-pyrrolic impurities) and chromatographed on neutral alumina. This separates chlorin and other impurities. On recrystallisation, purple lustrous crystals are left on the filter.[21] The yield is not large, about 20–25% (based on pyrrole), but the synthesis is simple and direct. Much porphyrin chemistry has been elucidated using this purple work-horse.

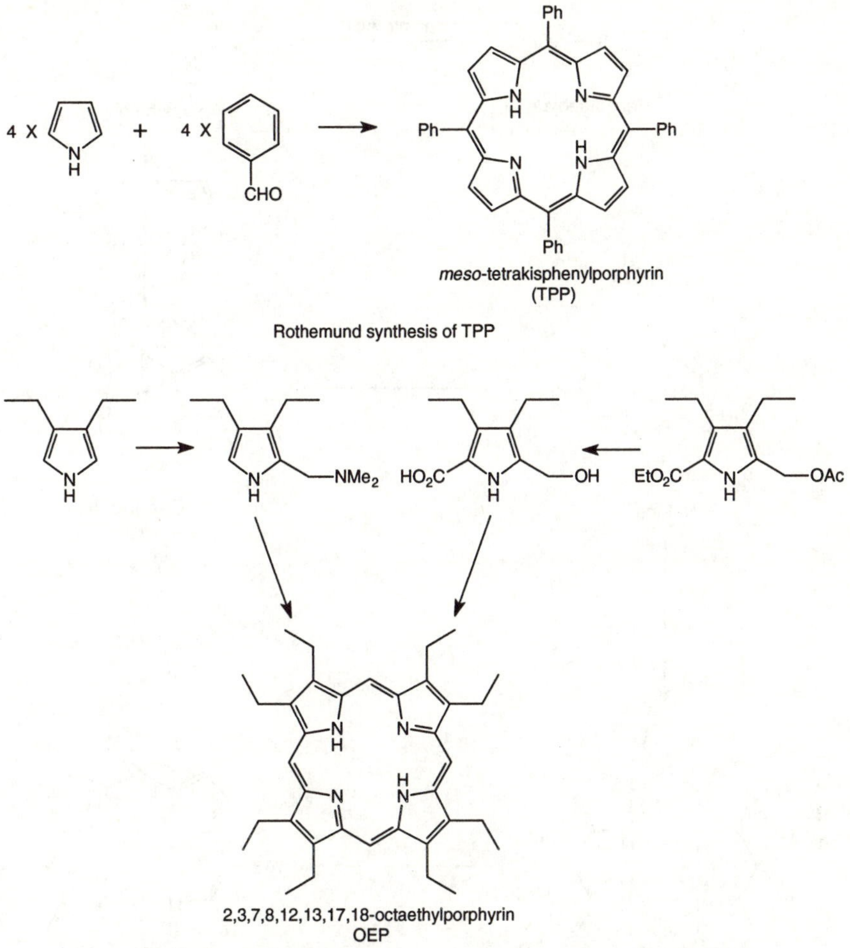

4 X
+
4 X
CHO
Ph
Ph
Ph
Ph
meso-tetrakisphenylporphyrin
(TPP)
Rothemund synthesis of TPP
NMe2
HO2C
OH
EtO2C
OAc
2,3,7,8,12,13,17,18-octaethylporphyrin
OEP

Fig. 2.25 Synthesis of TPP and OEP from monopyrroles.

The mechanism of this reaction is thought to involve formation of a carbonium ion by attack of a protonated aldehyde on a pyrrole α-position. The carbonium ion then goes on to attack the α-position on another pyrrole to give a *meso*-substituted dipyrrlymethane. Chain building continues until tetrapyrrlycarbinols (in various stages of oxidation) and black poly-pyrrolic by-products are formed. Ring closure follows to give porphyrinogens, which oxidise in air to porphyrins, and phlorins, which tautomerise to chlorins and oxidise slowly to porphyrins.

β-Octasubstituted porphyrins have been synthesised by heating 3,4-disubstituted pyrroles (which may or may not be further substituted in the α-positions) either dry or in solution. Fischer, for example, prepared

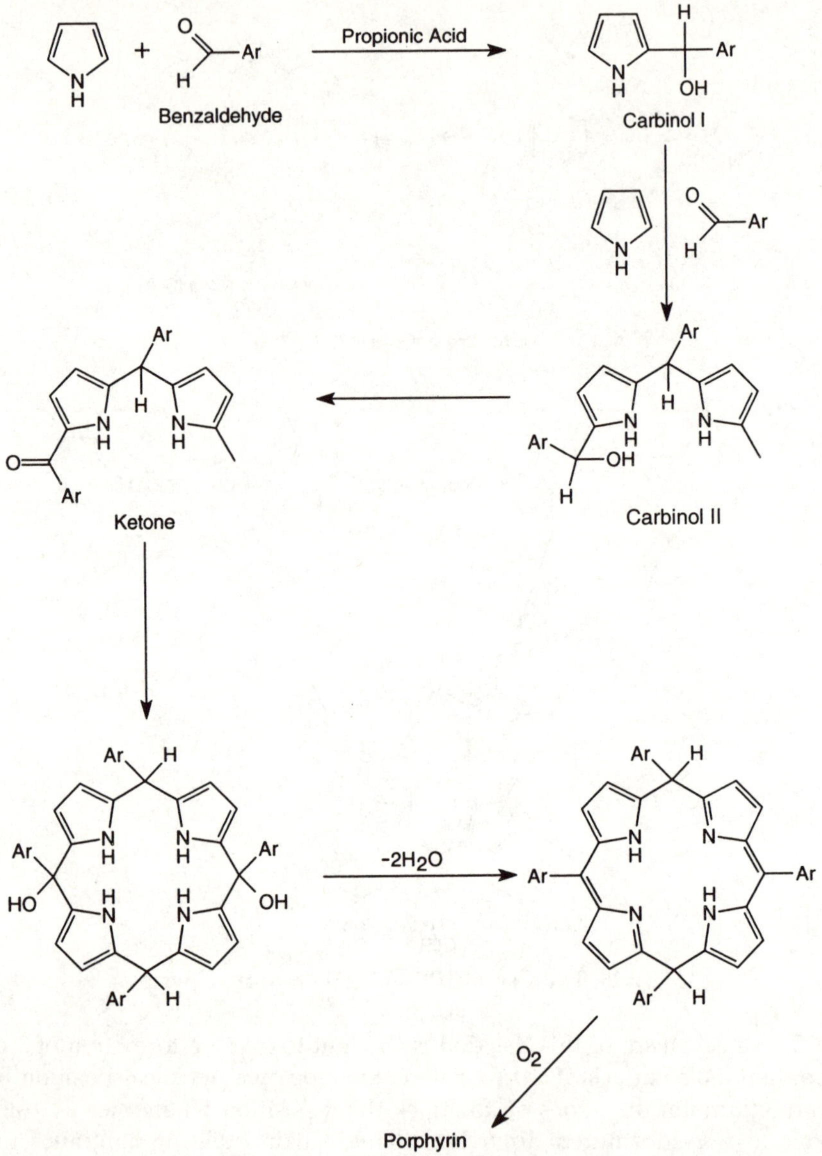

Fig. 2.26 Mechanism of TPP synthesis.

octamethylporphyrin by heating 3,4-dimethyl pyrrole in formic acid. Octaethylporphyrin is usually prepared by first treating 3,4-diethylpyrrole with dimethylamine and formaldehyde, to give the 2-*N*,*N*-dimethyl-aminomethylpyrrole, in a Mannich reaction. This product is then heated in refluxing acetic acid.[22] However, even this synthesis requires an experi-

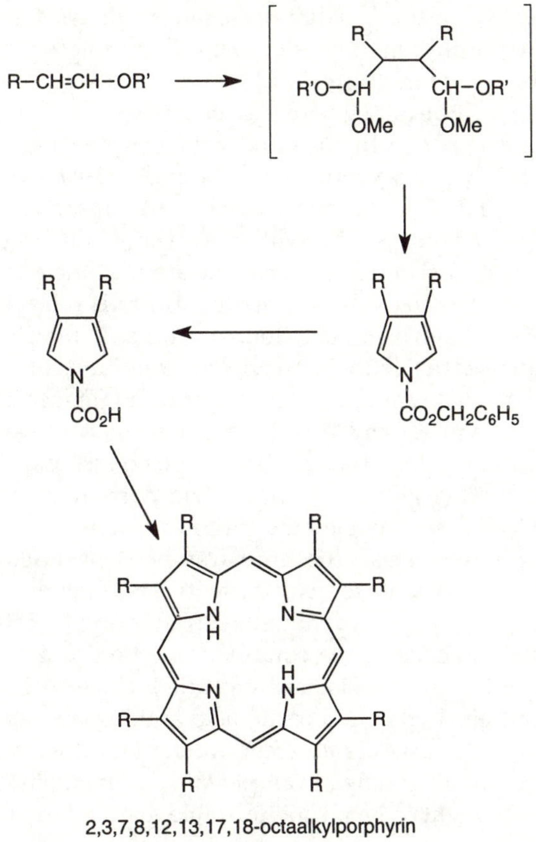

2,3,7,8,12,13,17,18-octaalkylporphyrin

Fig. 2.27 General synthesis of octaalkylporphyrins.

enced synthetic chemist and is not a trivial exercise. Yields of between 50 and 90% have been claimed. Professor H.H. Inhoffen, who first perfected the synthesis of this important porphyrin back in the 1960s, was a one-man porphyrin factory for many years. He was often mentioned at the end of papers for "his kind gift of OEP". Other routes to octaalkylporphyrins have been perfected in the last few years. Some examples[23] are shown schematically in Figure 2.28.

2.3.4 Porphyrins from dipyrrolic precursors

The three main types of dipyrrolic precursors are dipyrromethenes, dipyrrylmethanes, and dipyrrylketones, all of which have to be made from monopyrrolic precursors. This imposes symmetry restrictions on the types of porphyrins that can be made by these routes because they come together through condensation of the two dipyrrolic intermediates. Centrosymmetrically substituted porphyrins (and those in which one or both halves of the molecule are symmetric) are ideally prepared together via dipyrrolic precursors. These include all the naturally occurring porphyrins (e.g. protoporphyrin IX, and the type III uro- and coproporphyrins), which have a C–D dipyrrolic unit which is symmetrical about the 15-*meso*-carbon.

Hans Fischer and his Munich school are responsible for most dipyrromethene-based porphyrin syntheses. These can be substituted in their α-positions, e.g. with a reactive bromine atom at one end and a methyl group at the other. Alternatively, the methyl group is replaced with a bromomethyl group. Whatever α-substituents are used, the dipyrromethene usually self-condenses (using the quite violent conditions of an organic acid melt) to give the desired centrosymmetric porphyrin (Figure 2.28.)

For constructing type III and type IX porphyrins, two dipyrromethenes are used, one with bromine atoms in both α positions, and the other with methyl or bromomethyl groups. By using one pyrromethene that is completely symmetrical about the methine carbon atom, it is possible to avoid mixtures of porphyrins. This procedure was used by Fischer to prepare deuteroporphyrin IX, en route to his total synthesis of hemin.

The problem with using dipyrromethenes as dipyrrolic precursors for porphyrins, is the violence of the conditions needed to achieve coupling. Molten succinic or tartaric acid is hardly gentle but the success of Fischer's methodology (and his domination of the field and his laboratory) tended to stifle attempts to use or investigate other, milder (and therefore potentially more general) methodologies. For example, dipyrrylmethanes, as dipyrrolic precursors, were thought to be far too unstable, particularly to acids. Then,

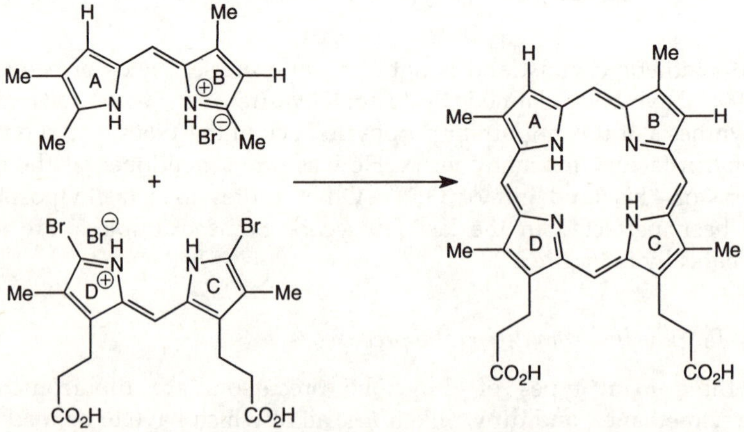

Fig. 2.28 Porphyrin synthesis from dipyrromethenes: Fischer's synthesis of deuteroporphyrin IX.

in 1960, MacDonald published his synthesis of porphyrins using a 5,5'-diformyldipyrrylmethane condensing with a 5,5'-di-unsubstituted dipyrrylmethane with the help of an acid (e.g. hydroiodic or toluene sulphonic acid) catalyst. About the same time, R.B. Woodward used a MacDonald approach for the total synthesis of chlorophyll.[24] The advantage of the MacDonald procedure is that being a milder synthetic technique it allows porphyrins with complex, more labile substituents to be made.

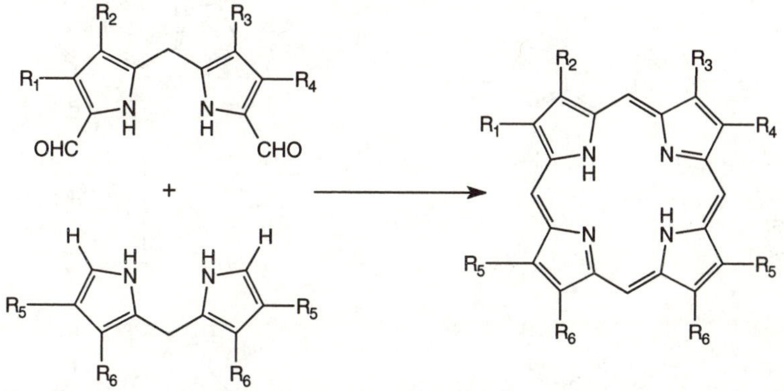

Fig. 2.29 Porphyrin synthesis from dipyrrylmethanes. The MacDonald method.

A variation of the MacDonald route is oxidation of readily available 5,5'-diformyldipyrrylmethanes, followed by condensation of the resulting 5,5'-diformyldipyrrylketone with 5,5'-di-unsubstituted dipyrrylmethanes (or 5,5'-dicarboxylic acids). This yields the corresponding oxophlorin (a porphyrin in which one *meso*-carbon is part of a ketone group, $>C=O$), which is reduced to the porphyrin by a variety of methods. Use of dipyrrylketones has one disadvantage in that 5,5'-di-unsubstituted dipyrrylketones are too unreactive, so that the ketone group has to be carried on the dipyrrole with the two formyl groups.

2.3.5 Porphyrins from open chain tetrapyrrolic intermediates

There is only one way to build up a porphyrin in a completely general way, and that is through an open-chain tetrapyrrolic intermediate, which is usually separated out prior to the final cyclisation step. Different types of tetrapyrrolic intermediates have been used; for example, bilanes, bilenes, oxobilanes, and biladienes. The fact that they are open-chain tetrapyrroles is inherent in the name; all these compounds refer in there own way to the bile pigments, the naturally occurring products of porphyrin catabolism.

According to the IUPAC rules, the term *bilane* is used to denote the unsubstituted open chain tetrapyrrole in which carbon atoms join together

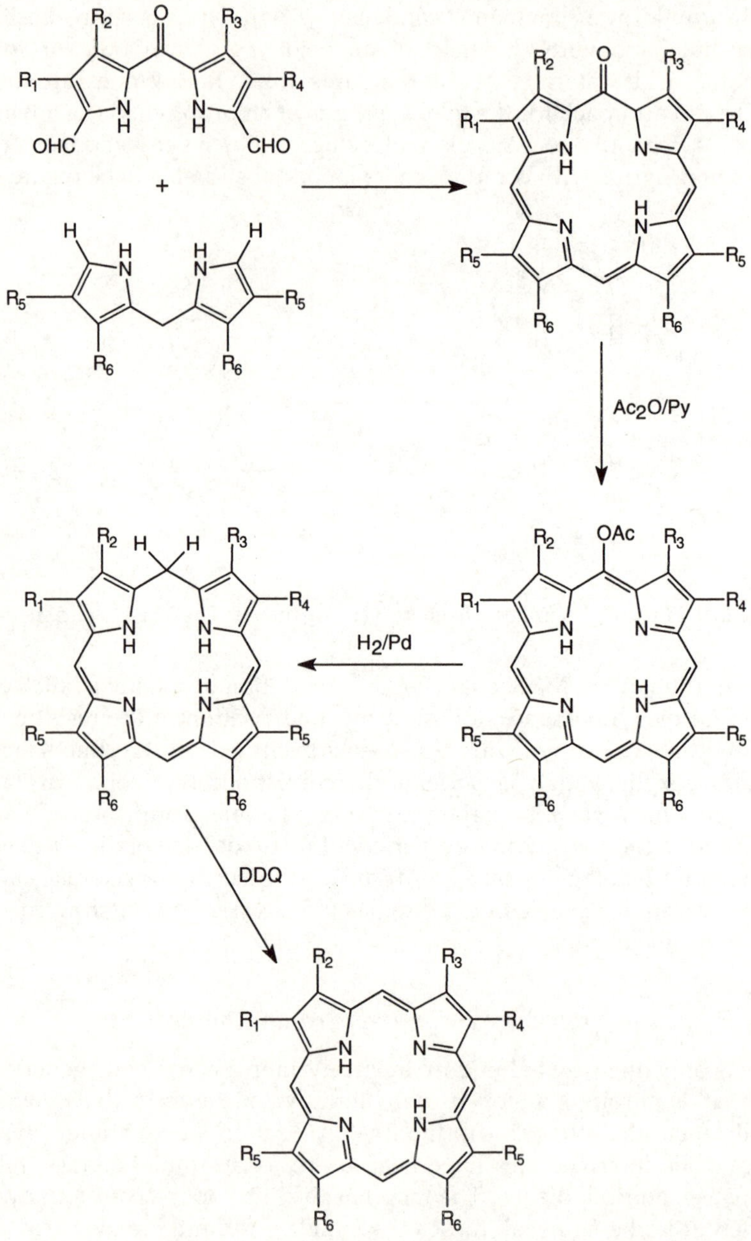

Fig. 2.30 Porphyrin synthesis from pyrroketones via oxophlorins.

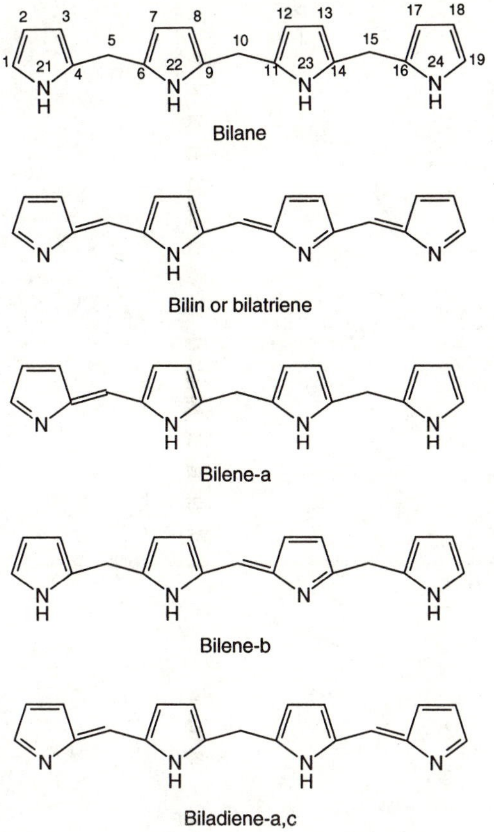

Bilane

Bilin or bilatriene

Bilene-a

Bilene-b

Biladiene-a,c

Fig. 2.31 Different types of open-chain tetrapyrroles.

the pyrrole units (the *meso*-carbons in ring-closed porphyrins). The numbering of the carbons is shown. Note how carbon number 20 (corresponding to the loss of the C-20 *meso*-carbon in porphyrins) is missed out to bring the numbering in line with the parent porphyrin system. As the carbon bridges become unsaturated, so the name reflects this (change of -ane to ene or -diene), along with the use of the lower case letters a, b, c to show which carbon bridge has been oxidised. When all three carbon bridges are oxidised, the name *bilin* is used. The interested reader is referred to the IUPAC rules on tetrapyrrole nomenclature, mentioned in Chapter 1 (see ref. 7).

The advantage of using such compounds in porphyrin synthesis is that, because of their construction, the substitution pattern of the final porphyrin is built in. It is possible to check this, using various spectroscopic techniques (e.g. mass spectroscopy and NMR) prior to cyclisation. Care has to be taken, however, in the use of these compounds, because the more

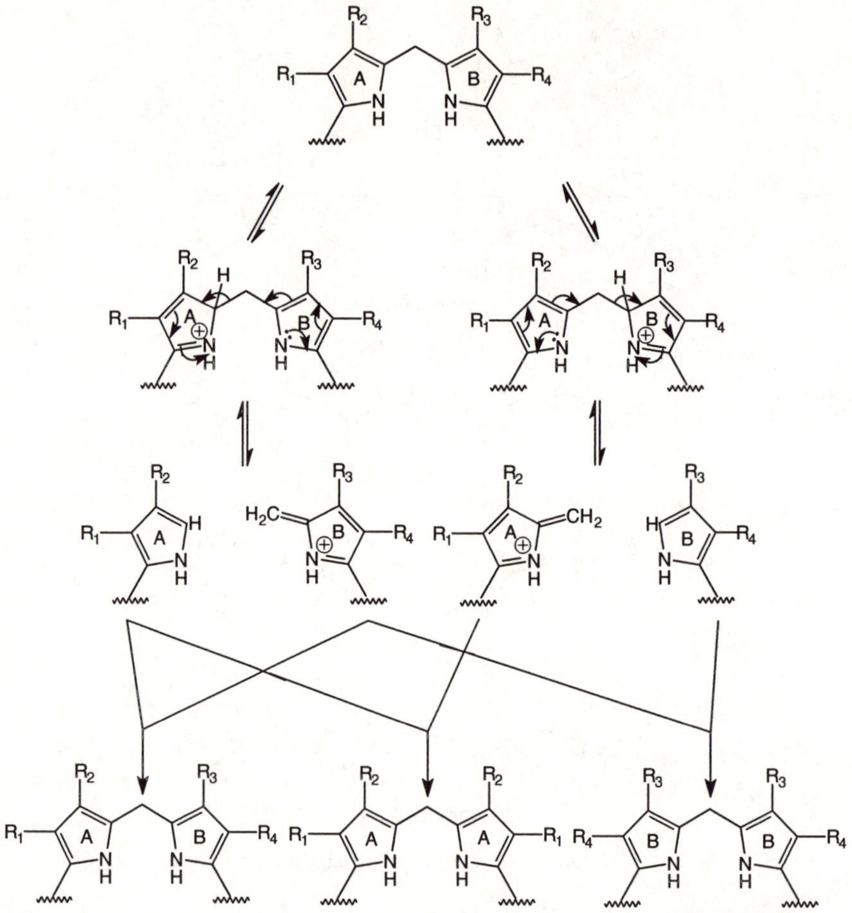

Fig. 2.32 Randomisation of bilanes in acid. Only rings A and B are shown but similar randomisation occurs for rings B and C, C and D, and D and A, leading to highly complex mixtures of bilanes.

saturated the open-chain tetrapyrrole (e.g. bilanes and bilenes), the more labile these compounds are to acid, and the easier it is for the different pyrrolic rings to become randomised. This leads to a maximum of porphyrins. Therefore, cyclisation conditions have to be extremely mild, or electron-withdrawing substituents must be present, to lower the reactivity of the open-chain tetrapyrrole. Consequently, the best routes to porphyrins involve oxobilanes and biladienes, where an electron-withdrawing group and unsaturation, respectively, stabilise the open-chain tetrapyrrole.

There are two routes involving oxobilanes, called a-oxobilane and b-oxobilane respectively, depending on which of the bridging carbon atoms

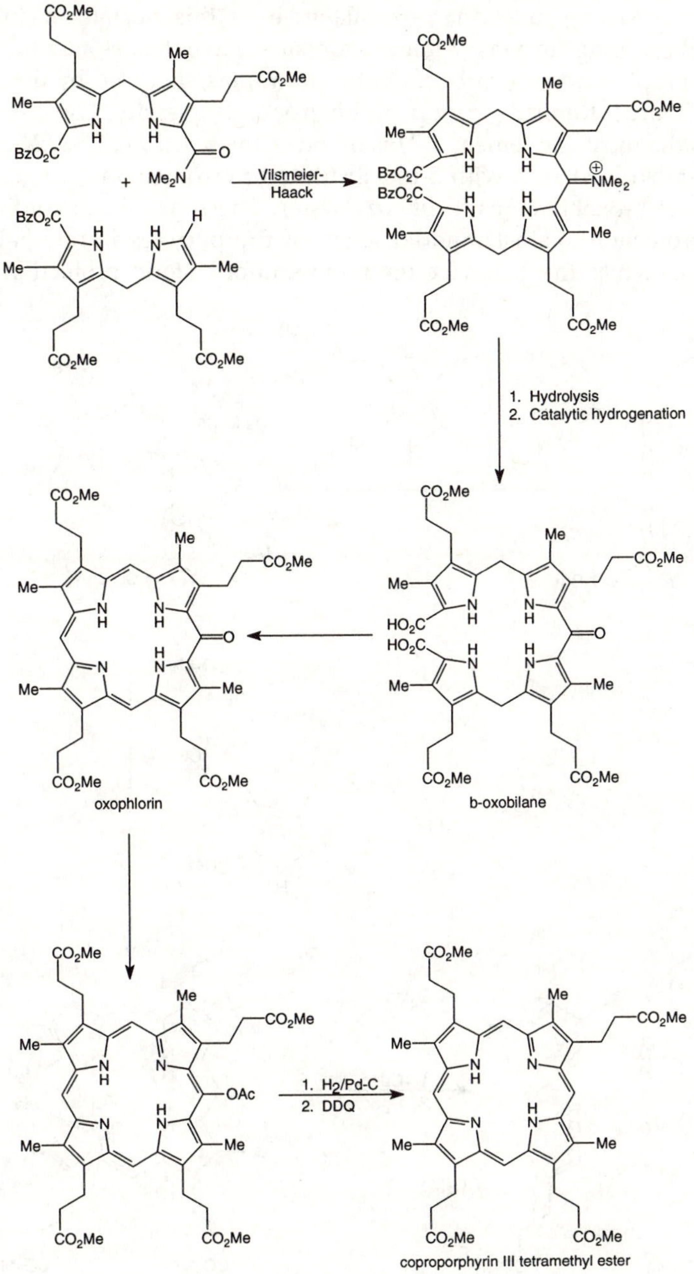

Fig. 2.33 Porphyrin synthesis via a b-oxobilane intermediate.

carries the ketone function: a-oxobilanes have this carbon on one of the terminal bridging carbons, while b-oxobilanes have the ketone functionality on the middle bridging carbon. Of the two routes, invented by the late Professor George Kenner's group at Liverpool University, the b-oxobilane route is the most convenient.[25] This involves the reaction of 5-*N,N*-dimethylamido-pyrromethanes with 5-unsubstituted pyrromethanes to give, eventually, the b-oxobilane as a dibenzyl ester. The b-oxobilane configuration gives protection against randomisation of the pyrroles in one half of the molecule, while the benzyl ester groups afford steric protection against

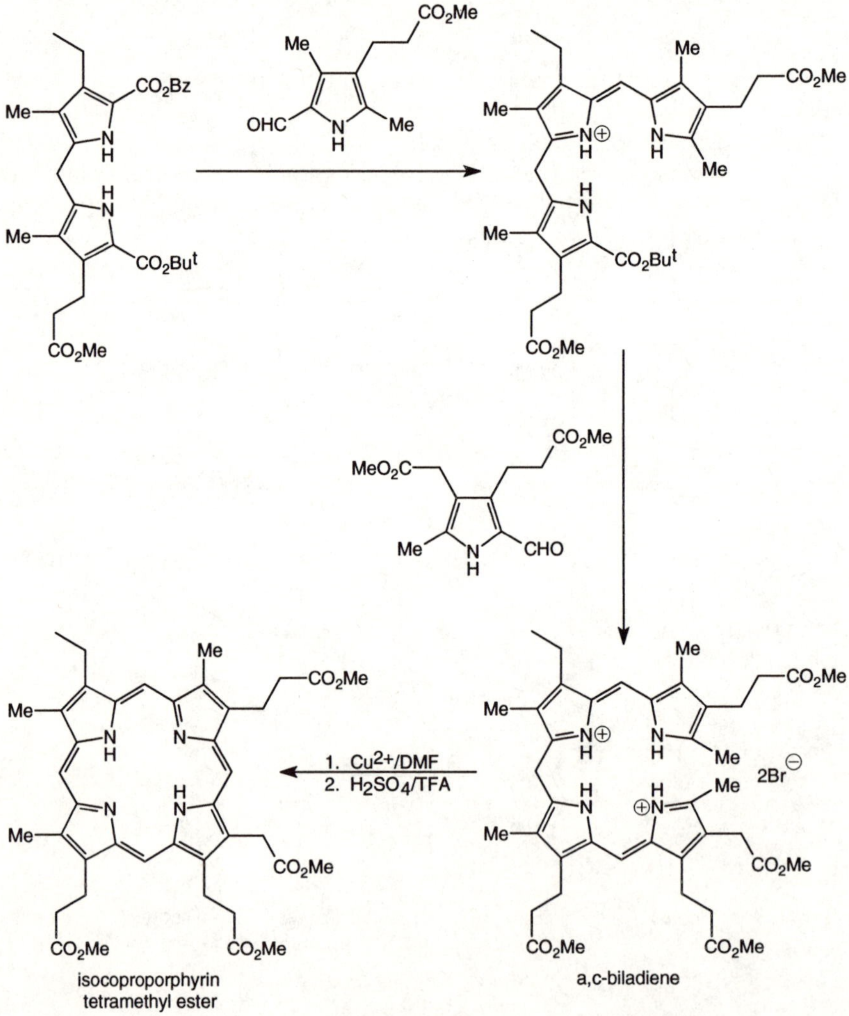

Fig. 2.34 Porphyrin synthesis via a,c-biladienes.

randomisation to the other half. The b-oxobilane can be cyclised to an oxophlorin directly without the necessity of taking out the oxo group first (as with a-oxobilanes). The oxophlorin is then reduced to the porphyrin. The synthesis of coproporphyrin III is given as an example.

There are several routes to a,c-biladienes. One was mentioned at the start of this section of porphyrin synthesis, i.e. condensation of two dipyrromethenes (one of which has an unsubstituted 5-position). Another route, to symmetrically substituted a,c-biladienes in particular, is by condensation of dipyrrylmethane-5,5'-dicarboxylic acids with two moles of a 2-formyl-5-pyrrole. This method can be used to construct the most unsymmetric porphyrins, e.g. isocoproporphyrin.[26]

Different methods of cyclisation of a,c-biladienes are used. In the example given at the beginning of this section, stannic chloride was used as a Friedel–Crafts catalyst to bolt the two ends of the molecule together. Another method is to use copper as a chelating agent, which causes the biladiene to wrap around it so bringing the ends of the molecule into close proximity. Reaction with sulphuric acid causes ring closure and demetallation of the copper.

2.4 References

1. See, F. Graham-Smith and B. Lovell in *Pathways to the Universe*, Cambridge University Press, Cambridge (1988), p. 94.
2. (a) S.L. Miller and L.E. Orgel, *The Origins of Life on the Earth*, Prentice–Hall, New Jersey (1974), p. 19; (b) L.E. Orgel, *The Origins of Life: Molecules and Natural Selection*, Chapman & Hall, London (1973), p. 100.
3. See ref. 2(a), p. 33.
4. See ref. 2(a), p. 83 and reference 2(b), p. 15.
5. See ref. 2(a), p. 96.
6. G. Ksander, G. Bold, R. Lattman, C. Lehmann, T. FrÅh, Y.-B. Xiang, K. Inomata, H.-P. Buser, J. Schreiber, E. Zass, and A. Eschenmoser, *Helv. Chim. Acta*, (1987), **70**, 1115.
7. A.R. Battersby, C.J.R. Fookes, G.W.J. Matcham, and E. McDonald, *Nature*, (1980), **285**, 17.
8. D. Shemin, *BioEssays*, (1989), **10**, 30.
9. C.G. Kannangara, S.P. Gough, P. Bruyant, J.K. Hooker, A. Kahn, and D. von Wettstein, *Trends Biochem. Sci.*, (1988), **18**, 139.
10. A.R. Battersby and F.J. Leeper, *Chem. Rev.*, (1990), **90**, 1261.
11. A.I. Scott, *Tetrahedron*, (1992), **48**, 2559.
12. L.J. Siepker, M. Ford, R. de Kock, and S. Kramer, *Biochem. Biophys. Acta*, (1987), **913**, 349.
13. S. Enosawa and A. Ohashi, *Biochem. Biophys. Res. Commun.*, (1986), **141**, 1145.
14. D.W. Nicholson, H. Kîhlr, and W. Neupert, *Eur. J. Biochem.* (1987), **164**, 147.
15. F.J. Leeper, *Nat. Prod. Rep.*, (1989), **6**, 171.
16. A. Eschenmoser, *Angew. Chem., Int. Edn. Eng.*, (1988), **27**, 5.

17. (a) D. Thibault, L. Debussche, D. FrÇchet, F. Herman, M. Vuilhorgne, and F. Blanche, *J. Chem. Soc., Chem. Commun.*, (1993), 513; (b) D. Thibaut, F. Kiuchi, L. Debussche, F.J. Leeper, F. Blanche, and A.R. Battersby, *J. Chem. Soc., Chem. Commun.*, (1992), 139; (c) A.R. Battersby, *Acc. Chem. Res.*, (1993), **26**, 15; (d) A.R. Battersby, *Science*, (1994), **264**, 1551.

18. (a) See K.M. Smith, in *Porphyrins and Metalloporphyrins*, ed. K.M. Smith, Elsevier Scientific Publishing, Amsterdam (1975), p. 29, (out of print), (b) D. Dolphin (ed.), *The Porphyrins*, Academic Press, New York (1978), Vol. 1, pp. 85, 101, 235, 265.

19. R.L.N. Harris, A.W. Johnson, and I.T. Kay, *J. Chem. Soc. C.* (1966), 22.

20. See ref. 18(b), p. 290.

21. A.D. Adler, F.R. Longo, J.D. Finarelli, J. Goldmacher, J. Assour, and L. Korsakoff, *J. Org. Chem.*, (1967), **32**, 476.

22. H.H. Inhoffen, J.-H. FÅhrop, H. Voigt, and H. Brockmann, *Ann. Chem.*, (1966), **695**, 133.

23. K.S. Chamberlain and E. LeGoff, *Heterocycles*, (1979), **12**, 1567; H.C. Callot, A. Louati, and M. Gross, *Angew. Chem., Int. Ed. Engl.*, (1982), **21**, 285.

24. R.B. Woodward, *Angew. Chem.*, (1960), **72**, 651; *Pure Appl. Chem.*, (1961), **2**, 383.

25. See ref. 18(a), p. 40 and ref. 18(b), p. 274.

26. A.W. Johnson and I.T. Kay, *J. Chem. Soc.*, (1965), 1620; A.R. Battersby, G.L. Hodgson, M. Ihara, E. McDonald, and J. Sanders, *J. Chem. Soc., Perkin Trans. 1*, (1973), 2923.

3. How do they do it?—Making oxygen

....

3.1 The electronic structure of porphyrins

3.1.1 Introduction

Why is grass green and blood red? Put another way, why is chlorophyll photoactive, beginning the process of photosynthesis, while the closely related heme performs a whole range of transport and redox functions that are non-photochemical in origin and depend on the protein that it is embedded in? To begin to answer these questions we need to know something about the chemical and physical properties of porphyrins. To understand those properties, we will require a knowledge of porphyrin electronic structure. That electronic structure is based on a skeleton of 20 carbon atoms surrounding a central core of four nitrogen atoms. This atomic arrangement supports a highly stable configuration of single and double bonds, called an aromatic π-system. The simplest and, indeed, archetypal aromatic π-system is seen in benzene.

3.1.2 The strange case of benzene

When organic chemists say that a compound is "aromatic", they are describing that compound's structure and properties. When August Kekulé first coined the term in 1865, to describe the peculiar stability of benzene and its derivatives, he could hardly have guessed that 130 years later his concept would still be fascinating chemists world-wide.

The first so-called aromatic compounds to be studied seriously, such as vanillin (derived from vanilla), had two obvious properties. They had a sweet smell and were remarkably stable. This last property was the reef on which many of the early theories of chemical bonding foundered. Consider benzene. Kekulé knew that its molecular formula was C_6H_6. The only way he could rationalise this formula with the known properties of benzene, was to imagine the six carbon atoms joined in a ring and connected by three alternate double bonds. This is where the trouble started because double bonds are supposed to confer reactivity on an organic molecule: benzene is stable. Double bonds can readily be added to; for example, they will undergo fast reactions with bromine and sulphuric acid to give simple "addition" compounds. The reagents simply "add" across the double bond,

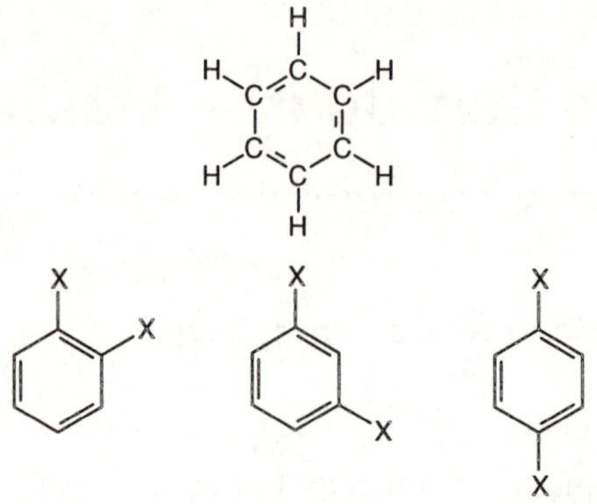

Fig. 3.1 Kekulé's benzene and three disubstituted benzene isomers.

destroying one half of it, leaving a single bond between the carbon atoms. Consequently, benzene with its three alternating double bonds, should be a real chemical performer: this is not so. Benzene resists such simple addition reactions and reacts with bromine and sulphuric acid only with difficulty; and then only in such a way as to maintain the integrity of its circular system of alternating double bonds. Somehow, this cyclic π-system seems to possess a special stability. Kekulé's initial structure for benzene did not answer this conundrum, but it did account for the number and nature of all the substituted benzene derivatives.

This structure, however, was flawed. For example, if there really were three alternating double bonds, then the benzene hexagon should be severely distorted. This is because double bonds are shorter than single bonds. In fact, the benzene hexagon is perfectly regular. Kekulé's critics scoffed and proceeded to publish alternative structures that to us now, make little chemical sense.

Then, Kekulé had his famous dream of snakes swallowing their own tales and half a dozen monkeys in a ring, doing some equally dubious things with theirs. Out of all this, Kekulé's intuition distilled a dynamic picture of the benzene molecule in which each carbon–carbon link oscillates between being a single and a double bond. This representation captured the imagination of most chemists and, with some modifications, survived well into the twentieth century. However, it was not until the advent of quantum mechanics and theories of chemical bonding that invoked molecular orbitals, that the real solution to the benzene problem was forthcoming, thanks to a German chemist, Erich Hückel. He developed a simple mathematical theory that dealt solely with the π-electrons in a molecule. It was so simple that even organic chemists (who as a species have tended to be mathematically illiterate) could understand it.[1]

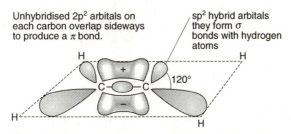

Unhybridised 2p² arbitals on each carbon overlap sideways to produce a π bond.

sp² hybrid arbitals they form σ bonds with hydrogen atoms

Fig. 3.2 sp^2 hybridisation of carbon in ethene, C_2H_4.

A double bond between, say, two carbon atoms, consists of two parts—a sigma (σ) molecular orbital of two tightly bound electrons, and a π molecular orbital of two more loosely bound (and therefore more energetic) electrons (see Figure 3.2). It is these π-electrons that make a double bond reactive.

In a cyclic molecule with alternating double bonds, such as benzene, the σ-electrons provide a framework for the π-bonds, and it is the geometry of this framework that is one of the factors determining whether the π-electrons will be "aromatically" stabilised. If the ring is flat, then aromatic stabilisation can take place.[2] The reason for this is that in a planar configuration, the three double bonds of benzene are no longer isolated, but delocalised. In other words, instead of six π-electrons forming three isolated π-bonds, they spread out around the ring giving two doughnut-shaped clouds of electron density above and below the six-carbon hexagonal framework. Hückel showed that by doing this, the π-electrons are energetically more stable than if they were isolated in three double bonds.

Not all cyclic molecules with alternating double bonds are stabilised in this way. For example, cyclooctatetraene has four alternating double bonds

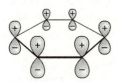

Six 2p atomic orbitals of hexagonal benzene make three π-bonding and three π^*-antibonding molecular orbitals

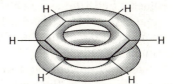

This is the lowest energy π-bonding molecular orbital. It is as though two "doughnuts" of π-electron probability waves sandwich the flat hexagonal C_6H_6 ring. This orbital contains two electrons.

Fig. 3.3 Delocalisation, in benzene, of two of its π-electrons. The 'aromatic' nature of benzene is achieved with the optimum flat geometry.

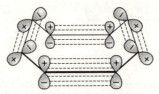

Fig. 3.4 Isolated, localised π-electrons of cyclooctatetraene. This molecule is highly unstable if flat (antiaromatic). The boat form is far more stable and is non-aromatic.

in its eight-sided ring, which behave as if they are isolated from each other. This molecule is highly reactive, without a hint of the stability of its six-sided counterpart, benzene.

Huckel's great contribution to the understanding of aromaticity was to predict exactly which cyclic molecules with alternating double bonds would be stable, and which would not. He showed for example that six (the number of π-electrons in benzene) was not the only magic number to confer aromatic stability. In 1931, Hückel announced that to be aromatic, a single-ring, planar compound had to have $[4n+2]$ π-electrons, where n is a whole number. So, a ring containing 2,6,10,14,**18** . . . $[4n+2]$ π-electrons may be aromatic (the reason why 18 is highlighted will become clear). If the ring contains $4n$ π-electrons, then there will be no aromatic stability. In fact, Hückel predicts that such molecules will be destabilised, or, as he put it, antiaromatic.

From this point on, organic chemists began to have a field day, as they tried (and succeeded) in synthesising all the possible permutations of cyclic, π-bonded molecules to see if Hückel had got his sums right. And between certain limits, he had. Thus, all the $4n$ systems turn out to be highly reactive compounds, many of which are chronically unstable in air. Similarly, the $[4n+2]$ systems have some degree of aromatic stability, depending on the geometry and ring size of the molecule. As the ring gets bigger, Hückel's simple "$[4n+2]$ rule" begins to break down. But it is the geometry that is important. Take away the planarity of these systems, and they lose their aromatic stability.

What is so special about the $[4n+2]$ formula, the aromatic equivalent of the stable octet of the periodic table? Let us take benzene, the aromatic molecule *par excellence*, as our model. Each of the six carbon atoms in the benzene ring donates a single π-electron to the whole π-system, making a total of six π-electrons. On the simplest view, the six π-atomic obitals combine to give six π-molecular orbitals, three bonding and three antibonding. Hückel's theory predicts that one of the bonding molecular orbitals is of lowest energy, while the other two have higher, but identical energies. These last two are said to be degenerate. Each bonding molecular orbital can take a maximum of two electrons, so that all the bonding molecular

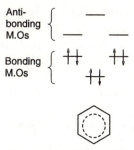

Fig. 3.5 Three bonding and three antibonding orbitals in benzene. Electrons are paired off and the system is aromatic.

orbitals are filled, i.e. $3 \times 2 = 6$. The other three molecular orbitals are all antibonding, which means that if electrons enter these orbitals, then the binding energy holding the molecule together decreases. The way that they are arranged is an energetic mirror image of the bonding molecular orbitals, i.e. the bottom two molecular orbitals are degenerate, while the highest energy antibonding molecular orbital is on its own. When benzene is in its ground electronic state, all the antibonding orbitals are empty. When ultra-violet light is shone into benzene, it excites an electron from the highest occupied molecular orbital into the lowest unoccupied antibonding orbital. It follows, therefore, that in an electronically excited state benzene has less binding energy and is chemically more reactive.

This energy pattern of molecular orbitals is repeated in other flat, cyclic systems with alternating double bonds. Here lies the reason for the differences in stability between aromatic [$4n+2$] and antiaromatic $4n$ systems. When [$4n+2$] electrons fill such a pattern of molecular orbitals, the electrons always pair up. This is a particularly stable arrangement. On the other hand, when $4n$ electrons try to fill the molecular orbitals, there are always

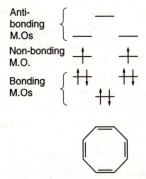

Fig. 3.6 In cyclooctatetraene, if it were flat, there would be two unpaired electrons in two non-bonding orbitals. This is unstable, i.e. antiaromatic.

two over, which go, singly, into a pair of degenerate molecular orbitals. So there are two unpaired electrons (called a biradical), which is a highly reactive situation. In the case of cyclooctatetraene, this molecule would rather be non-planar, with isolated double bonds, than flat, and have to exist as an even more reactive biradical.

Porphyrins are also aromatic compounds. The total number of π-electrons adds up to 22 (if $4n+2 = 22$, then $n = 5$), but only 18 of these are used to form the π-electron "doughnuts" above and below he porphyrin macrocycle.

We have dealt with the theoretical side of aromaticity, but what does it mean in practice? What peculiar properties of aromatic compounds do porphyrins partake of that allow them to do what they do? Once again, we shall consider benzene as the model aromatic compound and then see how its properties appear, modified, in porphyrins.

3.1.3 *Properties of aromatic compounds*

3.1.3.1 Electrophilic substitution   This is not meant to be a definitive account of aromaticity. It will suffice if we consider only one type of chemical reaction and two of the physical properties of benzene to demonstrate the point.

Aromatic compounds like benzene undergo a highly characteristic reaction called electrophilic substitution.[3] For example, halogens, such as chlorine and bromine, instead of simply adding to the formal double bonds as if it were an olefin (i.e. electrophilic addition in which both halogen atoms add to the double bond), displace one of the hydrogen atoms to give a monosubstituted aryl halide

What happens is that the halogen molecule approaches the benzene π-cloud of electrons, becomes polarised [i.e. one end becomes slightly positively charged (electrophilic) while the other end becomes slightly negatively charged (nucleophilic)]. Then, the electrophilic end of the halogen attacks one of the carbon atoms, leaving a positive charge on a carbon atom next to it. If this were an isolated double bond, then the halide anion that is formed would attack that positively charged carbon. However, the stability of the aromatic π-system ensures a different result. The halide anion removes a proton from the carbon atom with a halogen atom already attached, reforming the aromatic π-system and forming a hydrogen halide. The net result is that benzene plus halogen gives aryl halide plus hydrohalic acid. For chlorine, the reaction is slow but can be speeded up by catalysts, such as iodine or ferric ions.

There are many reactions of this general type, e.g. nitration (using the nitronium ion, NO_2^+, produced on mixing concentrated nitric and sulphuric acids), sulphonation (using sulphuric acid), and the industrially important Friedel–Crafts reaction. Here, molecules such as alkyl and acyl halides,

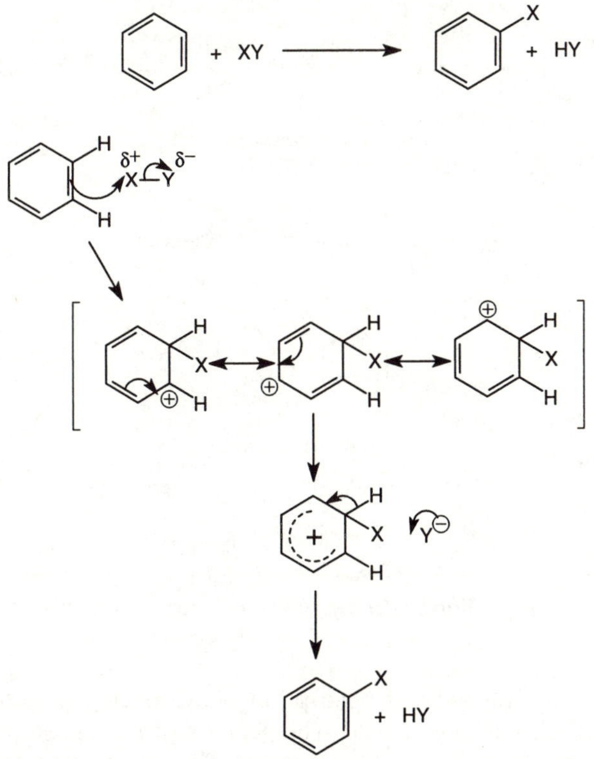

Fig. 3.7 Electrophilic substitution in benzene and its mechanism.

carbon monoxide, and oxygen, can be coaxed into reacting with benzene in the presence of aluminium trichloride, to give alkylbenzenes, carbonyl derivatives, benzaldehyde, and phenol, respectively.

The point about this type of reaction is that it demonstrates the lengths benzene and other aromatic molecules will go to in order to preserve their aromatic integrity.

3.1.3.2 NMR and UV–visible spectroscopy Aromaticity is directly observable using two powerful forms of spectroscopy. Nuclear magnetic resonance (NMR) spectroscopy depends on the nucleus of an atom having the property of spin (typically, the spin quantum number $= \frac{1}{2}$ in common atoms such as ^{1}H, ^{13}C, and ^{19}F, although other values are also possible). Such nuclei act like tiny magnets and become aligned when placed in a magnetic field. The field defines two nuclear energy levels, one aligned to the field and of lower energy, and the other opposed to the external field and of higher energy. A nucleus can be induced to flip its spin state, from aligned to opposed, by using radiofrequency (RF) radiation. The NMR spectrum

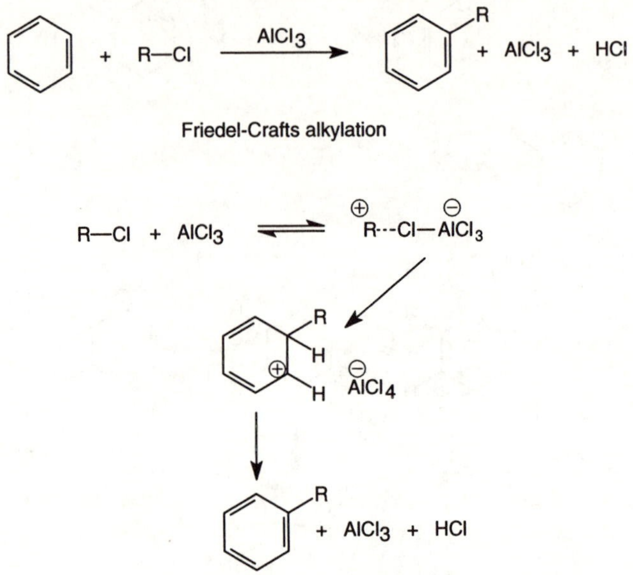

Friedel-Crafts alkylation

Fig. 3.8 Friedel–Crafts alkylation and its mechanism.

arises when the intensity of absorption of RF energy is plotted, against changing radiofrequency or field. The position of peak absorption depends on the immediate magnetic environment. This, in turn, results from the magnetic field generated by the electrons in other atoms surrounding the resonating nucleus in a molecule. It is as if this locally generated magnetic field diamagnetically shields the nucleus from the external magnetic field. The greater this diamagnetic shielding effect, the less the external magnetic field has to be to bring that particular atom into resonance. Differences in the field strengths at which signals are obtained for nuclei of the same kind, such as protons, but located in different molecular environments, are called chemical shifts. Also, at high resolution, NMR absorption lines split into multiplets, due to spin–spin coupling between groups of nuclei. NMR spectroscopy is therefore a very powerful tool for investigating molecular structure, because it can be used to investigate which functional groups of atoms are in the immediate environment of a resonating nucleus.

Now, benzene has six π-electrons delocalised over the six-carbon skeleton. In an external magnetic field, these electrons rotate around the hexagonal ring to give a diamagnetic ring current. This generates an induced magnetic field that strongly opposes the applied field *inside* the benzene ring but assists it *outside* the ring. The diamagnetic shielding effect is therefore very strong in benzene, so that the peaks due to the hydrogen atoms attached to the ring carbons are strongly shifted downfield. In other words,

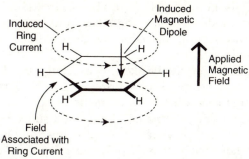

Fig. 3.9 Diagram showing the circulation of the π-electrons in an aromatic (to produce a diamagnetic ring current) under the influence of an applied magnetic field. The induced field opposes the applied field *inside* the ring but aligns with it *outside*.

the applied field needed to bring the benzene protons into resonance is much less than, say, for an aliphatic proton. This downfield shift in its NMR spectrum is a sure giveaway that a molecule has a ring current and is, therefore, aromatic. We shall see that such a ring current is particularly important in the porphyrin series.

In UV–visible spectroscopy, the absorbtion of ultraviolet photons by benzene causes an electron to be excited out of an occupied π-orbital into an unoccupied π*-orbital. Such a transition leads to vibrational and rotational changes in the molecule, so that the two absorbtion bands in the ultraviolet region (a strong band around 200 nm and a weaker one near 260 nm) are not narrow but broad. These bands are actually envelopes containing many vibrational transitions with rotational fine structure. The exact position and intensity of these bands is finely tuned by the substituents attached to the

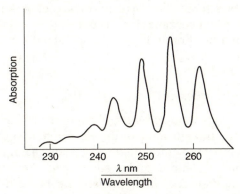

Fig. 3.10 UV absorption spectrum of benzene (in cyclohexane) showing the 'benzenoid' band.

benzene ring. Thus, groups such as vinyl (—CH=CH$_2$), hydroxyl (—OH), amino (NH$_2$), and iodo- (I) cause these bands to increase in intensity and shift to longer wavelength; that is to say, these groups are *auxochromes* of the benzene *chromophore*.

Under high resolution, the weaker lower energy band is found to be a composite of several narrow peaks (corresponding to vibrational fine structure) and is highly characteristic of aromatic compounds. No band like it exists for molecules with alternating double and single bonds that are not aromatic (i.e. conjugated polyenes). Therefore, this band is, not surprisingly called the benzenoid band. It is of only weak intensity because the electronic state of benzene corresponding to an electron excited out of a π-orbital into a π*-orbital has the same electronic symmetry as the ground state benzene. One of the rules (called selection rules) governing the way in which electrons are allowed to jump from one molecular orbital to another states that transitions between electronic structures of similar symmetries are forbidden. They become weakly allowed, because vibrations within the benzene molecule are different in the ground and excited states. The mixing of these vibrational states into the electronic states causes the symmetries of the benzene electronic ground and excited states to differ just enough for electronic transitions to be weakly allowed.

The two main absorption bands of benzene occur in the ultraviolet region. However, as aromatic rings increase in size (i.e. from six π-electrons to 10, 14, 18, etc.), or as benzene rings are fused together in a linear manner to make polynuclear aromatic hydrocarbons, such as naphthalene and anthracene, then less energy is required to excite an electron from a π-orbital into a π*-orbital. This means that the absorption bands are shifted to longer wavelength. Naphthacene, for example, has four benzene rings fused together, and the benzenoid band is shifted far enough to longer wavelengths to be in the visible region of the electromagnetic spectrum. Naphthacene is therefore yellow. The next member of the series, pentacene, with five linearly fused benzene rings, is blue.

We are now in a position to begin to understand some of the important properties of porphyrins.

3.1.4 *Properties of porphyrins*

3.1.4.1 Reactivity First and foremost, porphyrins are aromatic molecules. For example, they undergo some of the electrophilic substitution reactions characteristic of aromatic compounds—nitration, halogenation, sulphonation, formylation, acylation, and deuteration. Porphyrins differ from molecules such as benzene, in that there are *two* different sites on the macrocycle where electrophilic substitution can take place with different reactivities;[4] the *meso*-position and the pyrrole β-position. Which of these

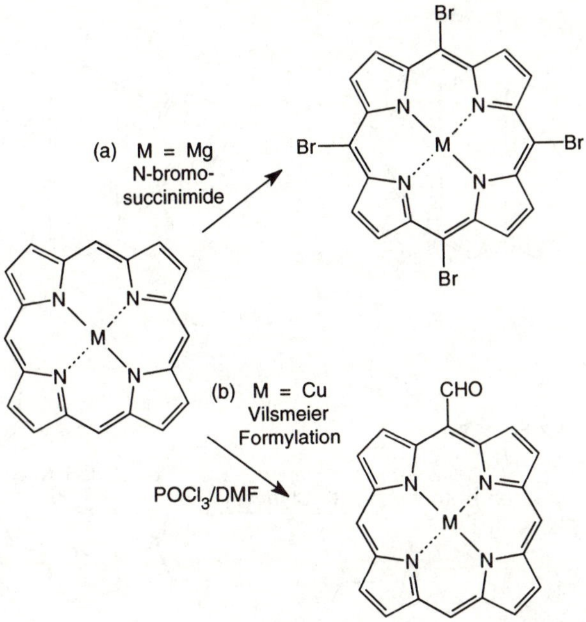

Fig. 3.11 (a) Bromination and (b) formylation of a metalloporphyrin.

sites is activated depends on how electronegative the porphyrin is. This can be controlled by the choice of metal to coordinate to the central nitrogen atoms. Thus, the introduction of divalent central metals produces electronegative porphyrin ligands (whose degree of electronegativity follows the order, MgP>ZnP>CuP>NiP>PdP), and these complexes are substituted on their *meso*-carbons. On the other hand, metals in electrophilic oxidation states (e.g. Sn^{IV}) or the free-base porphyrin (where formally, M = 2H) tend to deactivate the *meso*-carbons and activate the β-pyrrole positions to electrophilic attack. We can begin to get an insight into why these two sites have different reactivities to electrophilic substitution if we take a closer look at the canonical forms of the porphyrin.

Here we see that the macrocycle consists of two pyrrole units and two pyrrolenine units. After a π-electron count, the great American chemist, R.B. Woodward, pointed out that the two pyrrole units could be considered to have their own aromatic sextet of electrons. Each pyrrolenine unit, on the other hand, is one electron short of the aromatic sextet, having only five π-electrons. In order to make up this deficiency, they can be thought of as borrowing electron density from the neighbouring *meso*-carbons.[5] This means that, compared with the pyrrole carbons, the *meso*-carbons will have a tendency to be electron deficient and, therefore, would be less enthusiastic about electrophilic substitution.

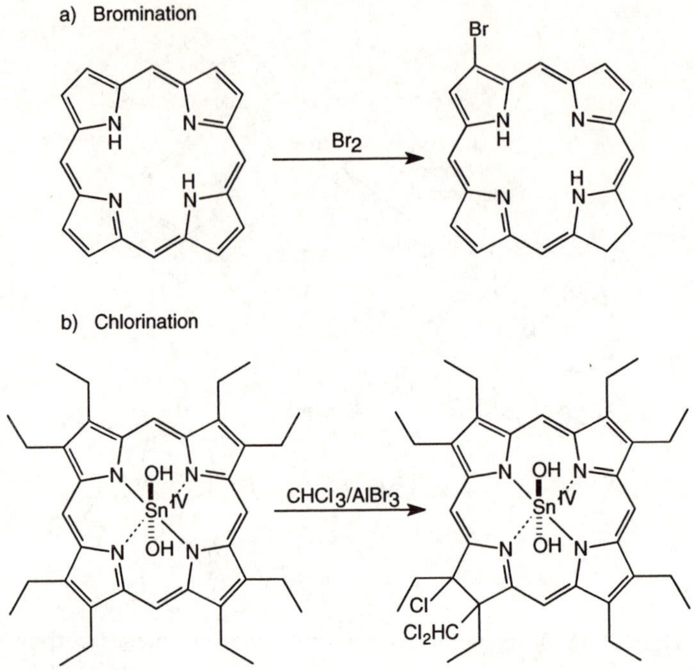

a) Bromination

b) Chlorination

Fig. 3.12 (a) Bromination of free-base porphyrin and (b) chlorination of $Sn^{IV}OEP$ with $CHCl_3/AlBr_3$.

There are, of course, exceptions to these simple rules. Thus, attempted formylation of free-base aetioporphyrin I under Vilsmeier conditions, leads to *meso*-monochlorination [Figure 3.14(a)], while chlorination of free-base octaethylporphyrin (either with mixtures of hydrochloric acid and hydrogen peroxide, or with sulphuryl chloride) leads to good yields of *meso*-5,10,15,20-tetrachloro-2,3,7,8,12,13,17,18-octaethylporphyrin [Figure 3.14(b)].

This last reaction shows that it is, in fact, difficult to monochlorinate porphyrins in their *meso*-positions. A way round this is to use the chlorin

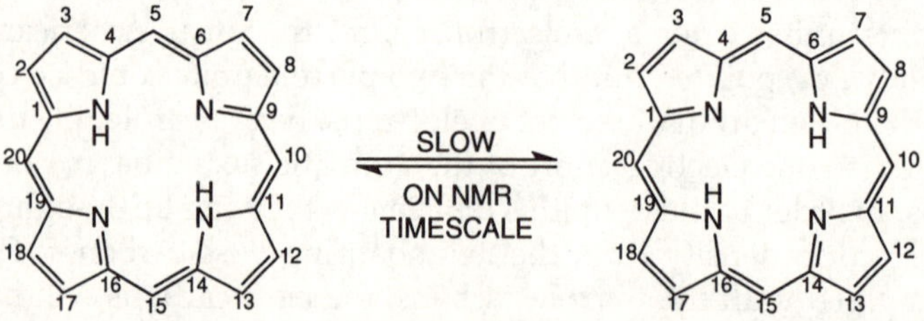

Fig. 3.13 The canonical forms of the porphyrin macrocycle. The form with hydrogen-bearing nitrogens adjacent to one another is ruled out on steric grounds.

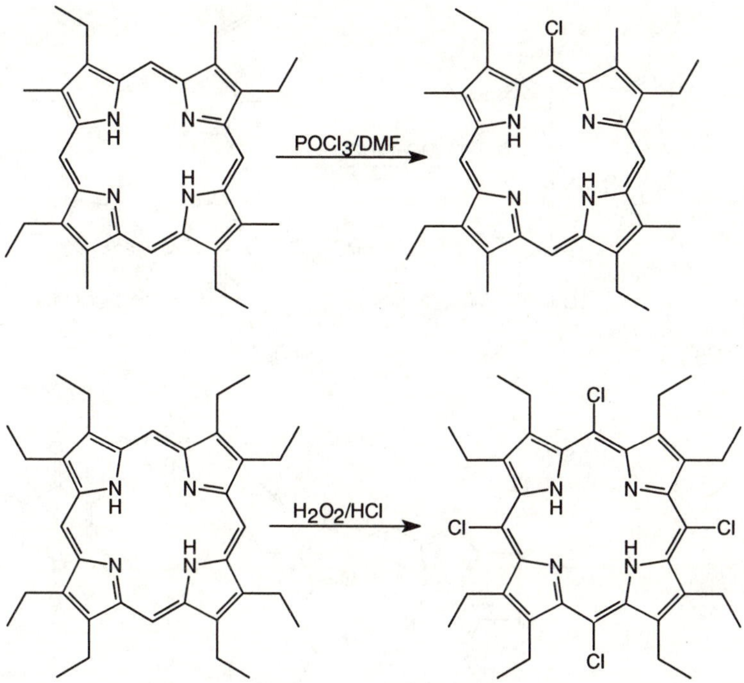

Fig. 3.14 *Meso*-chlorination of etioporphyrin I and OEP. No β-pyrrole chlorination occurs here.

instead of the porphyrin. It is known that *meso*-carbons next to reduced pyrrole rings are more susceptible to electrophilic attack than those next to ordinary pyrroles. Thus monochlorinated porphyrins can be made by chlorinating the chlorin and then oxidising the subsequent monochlorinated chlorin up to a porphyrin.

Deuteration at the *meso*-carbons (which is useful in biosynthetic studies) is achieved using free-base porphyrins and deuterated acids, or with metalloporphyrins (e.g. Mg) in pyridine solution and deuterated methanol. The general mechanism is thought to proceed via formation of isoporphyrin cations.

Like benzene derivatives, porphyrins will undergo Friedel–Crafts reactions. Thus, deuterohemin dimethyl ester acylates at its 3- and 8-pyrrole positions [Figure 3.17(a)]. Nitration [Figure 3.17(b)] occurs, unexpectedly, only at the *meso*-carbons, *even when the less electrophilic* β-pyrrole positions are vacant.

Electron-deficient species, such as carbenes and nitrenes, attack the porphyrin macrocycle, but in completely different ways. Thus, carbenes tend to attack at the β-pyrrole positions [Figure 3.18(a)] to give mainly cyclo-

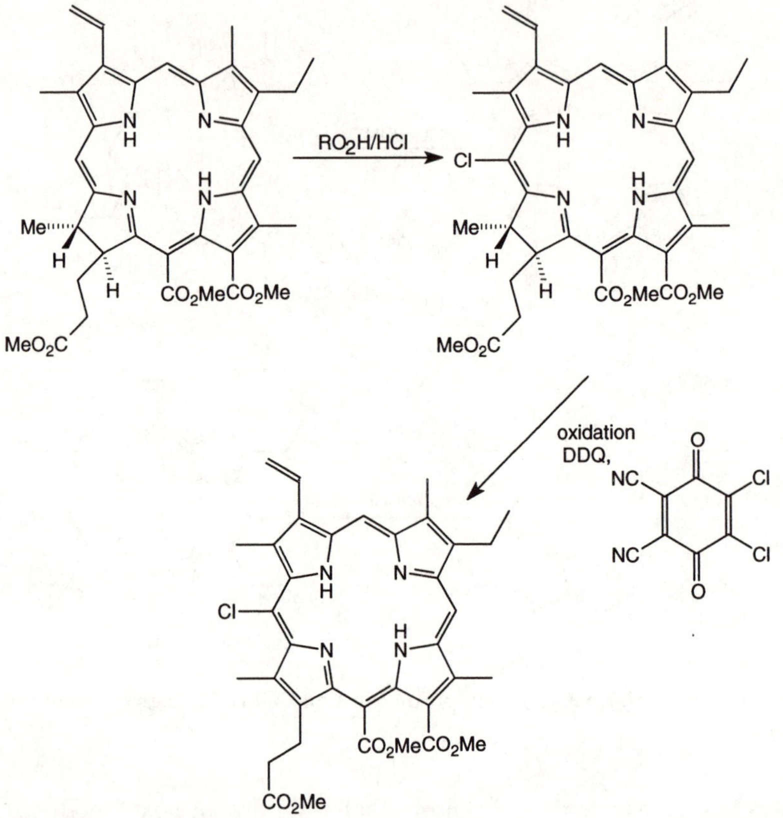

Fig. 3.15 Specific chlorination of a chlorin and oxidation to a *meso*-monochloro-porphyrin.

propane derivatives, while nitrenes attack at the *meso*-carbons to give ring-expanded compounds called homoazaporphyrins, which readily extrude the nitrogen atom on heating or metal chelation [Figure 3.18(b)].

Just as phenols are hydroxybenzenes, so porphyrins can be oxidised to yield hydroxyporphyrins. Thus, oxidation of zinc porphyrins with thallium(III) trifluoroacetate, followed by demetallation, yields an oxophlorin which is in tautomeric equilibrium with the *meso*-hydroxyporphyrin from. Uv–visible spectroscopy shows that, unlike phenols, the equilibrium is well on the side of the keto form, oxophlorin. Metals, on the other hand, tend to stabilise the *meso*-hydroxyporphyrin form (Figure 3.19).

The aromaticity of porphyrins is also indicated by measurements of their heats of combustion (because of aromatic stabilisation, benzene gives out less heat when it is burnt than if it consisted of three alternating double bonds, as cyclohexatriene). Also, X-ray crystallography of many porphyrins

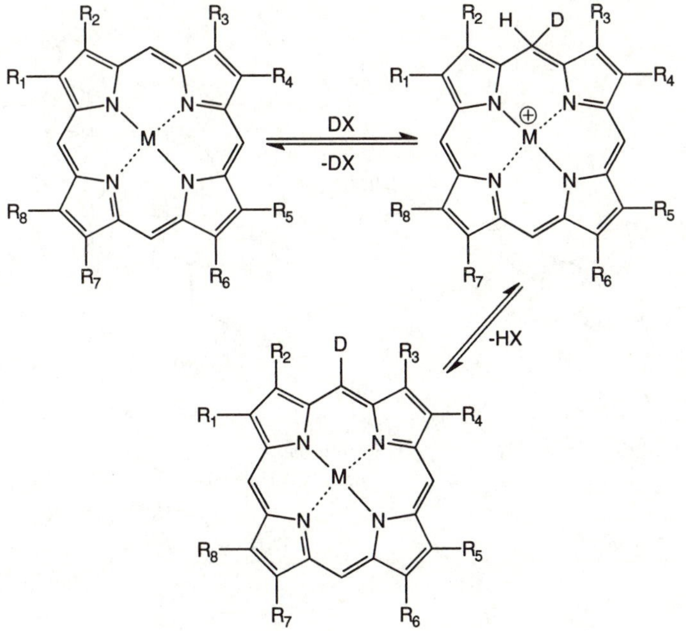

Fig. 3.16 Deuteration of porphyrins.

has shown the essentially planar topology of the porphyrin macrocycle (see Figures 1.5–1.8), a more or less basic prerequisite for aromaticity.

3.1.4.2 NMR spectra of porphyrins Perhaps the most telling piece of evidence that demonstrates the aromatic nature of the porphyrin macrocycle comes from NMR studies. These show that a diamagnetic ring current deshields the β-pyrrole protons and the *meso*-protons. Because the latter are attached to essentially electron-deficient carbons, they are shifted downfield further than the β-pyrrole protons. However, the inner N-H protons are shifted upfield, beyond tetramethylsilane (TMS, the arbitrary zero for NMR measurements in non-aqueous solutions). This is because the ring current, which deshields protons from the external magnetic field outside the macrocycle, shields them if they are inside the macrocycle (see Figure 3.9). This effect is not observed in benzene because it has no internal protons, but is in some larger aromatic molecules (e.g. 18-annulene), and under the right conditions these protons may be observed diamagnetically shielded and shifted upfield beyond TMS.

As an aromatic molecule, a π-electron count of the porphyrin macrocycle reveals 22 π-electrons, which conforms nicely with Huckel's [4n=2] rule for aromaticity ($n = 5$). However, only 18 π-electrons are considered to lie in

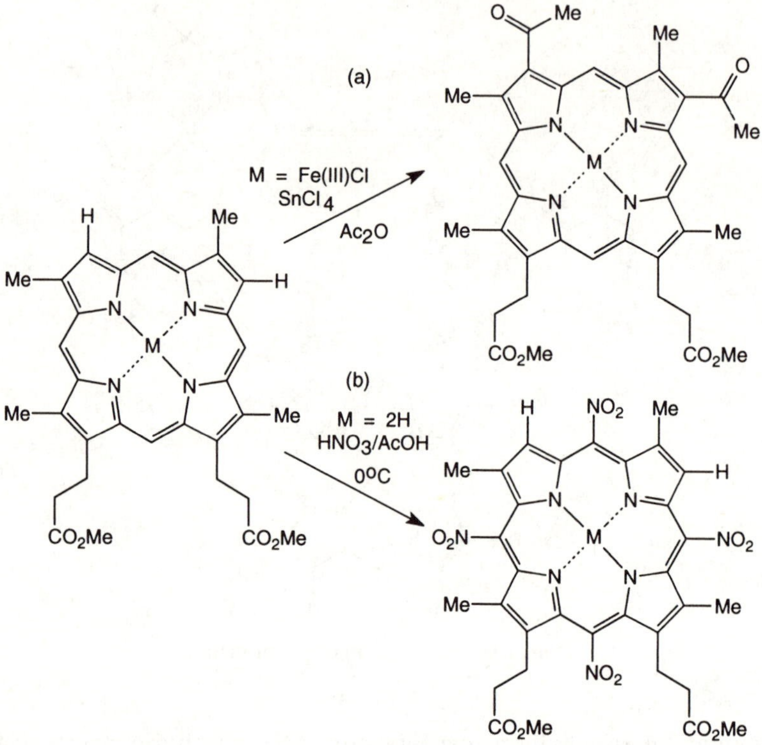

Fig. 3.17 (a) Acylation and (b) tetranitration of deuterohemin dimethyl ester.

the main delocalisation pathway, giving a [4n+2] aromatic system with n = 4 (as in 18-annulene, to which porphyrins bear more than a passing resemblance). In other words, four π-electrons are left out of the electron count. It is interesting that, when looked at this way, it is possible to explain the persistence of aromaticity in chlorins and bacteriochlorins, which only have a total of 20 and 18 π-electrons, respectively.

Thus, reduction of porphyrins under a variety of conditions gives chlorins in which two protons have been added across the β – β double bond of one of the pyrrole moieties (Figure 3.20), reducing it to a pyrroline ring. Chlorins are relatively stable entities (they can be oxidised back up to the porphyrin by high potential quinones, e.g. DDQ and chloranil) which are still aromatic systems; the reduction at the β-pyrrole carbons does not interrupt the aromatic delocalisation pathway. In fact, two more protons may be added across another β – β double bond (to give a mixture of tetrahydroporphyrins, or bacteriochlorins, see Figure 3.21) without disruption of the aromatic delocalisation pathway. Notice here that there are two possible tetrahydroporphyrins: a 2,3,7,8- and a 7,8,17,18-tetrahydroporphyrin.

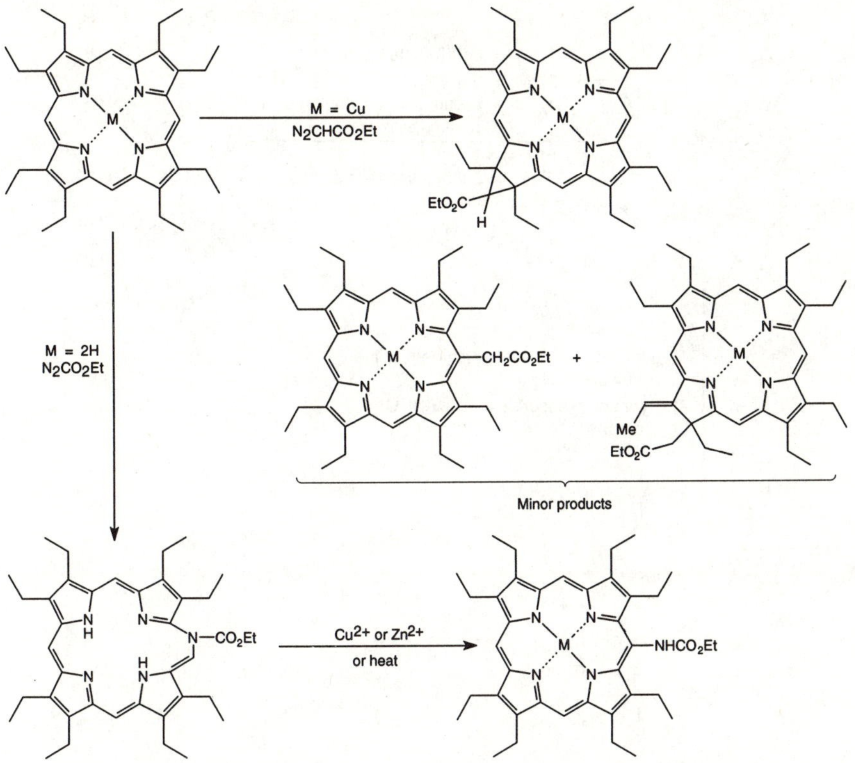

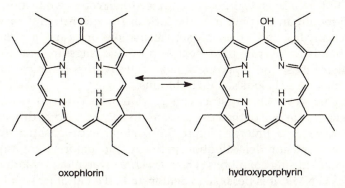

Fig. 3.18 Attack of carbene and nitrene on metallo-OEP.

oxophlorin   hydroxyporphyrin

Fig. 3.19 An oxophlorin and a 'hydroxyporphyrin'.

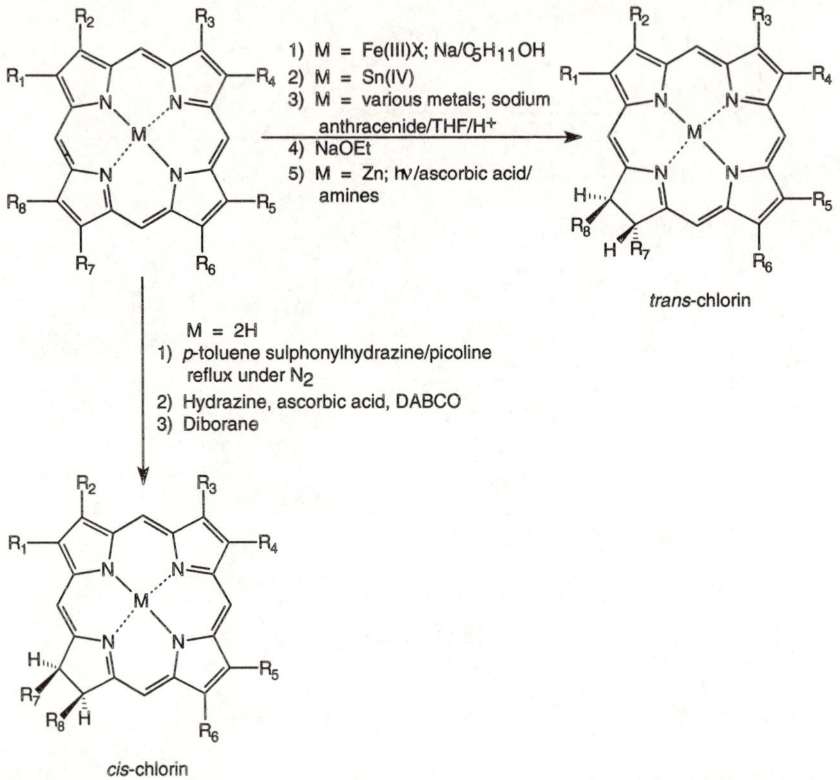

1) M = Fe(III)X; Na/C$_5$H$_{11}$OH
2) M = Sn(IV)
3) M = various metals; sodium anthracenide/THF/H$^+$
4) NaOEt
5) M = Zn; hν/ascorbic acid/ amines

trans-chlorin

M = 2H
1) *p*-toluene sulphonylhydrazine/picoline reflux under N$_2$
2) Hydrazine, ascorbic acid, DABCO
3) Diborane

cis-chlorin

Fig. 3.20 Reduction of porphyrins to give chlorins. None of these methods can be applied without protection of reducible side chains (e.g. vinyl). Also, there is no selectivity for which pyrrole ring is reduced unless there are bulky substituents on adjacent *meso*-carbons.

Which is produced depends on the steric constraints imposed on the reaction; bulky *meso*-substituents tending to favour production of the 7,8,17,18-tetrahydroporphyrin, while lack of steric constraints favours the 2,3,7,8-tetrahydro derivative. Model calculations indicate that the β – β double bonds on pyrrole units next to pyrroline rings are much more reactive than those on opposite rings. Tetrahydroporphyrins are more susceptible to aerial oxidation than chlorins.

An isomer of a chlorin (strictly, a 2,3-dihydroporphyrin) is formed by hydrogenation at one of the *meso*-carbons, to give a 5,22-dihydroporphyrin, better known as a phlorin (Figure 3.22.) Here, hydrogenation *has* interrupted the aromatic delocalisation pathway and phlorins are only stable under non-oxidising conditions. They *isomerise* reversibly to chlorins.

Using ^{13}C NMR, it is possible to demonstrate the tautomerism of the porphyrin N-H protons. At room temperature, the α-carbons (i.e. carbon atoms

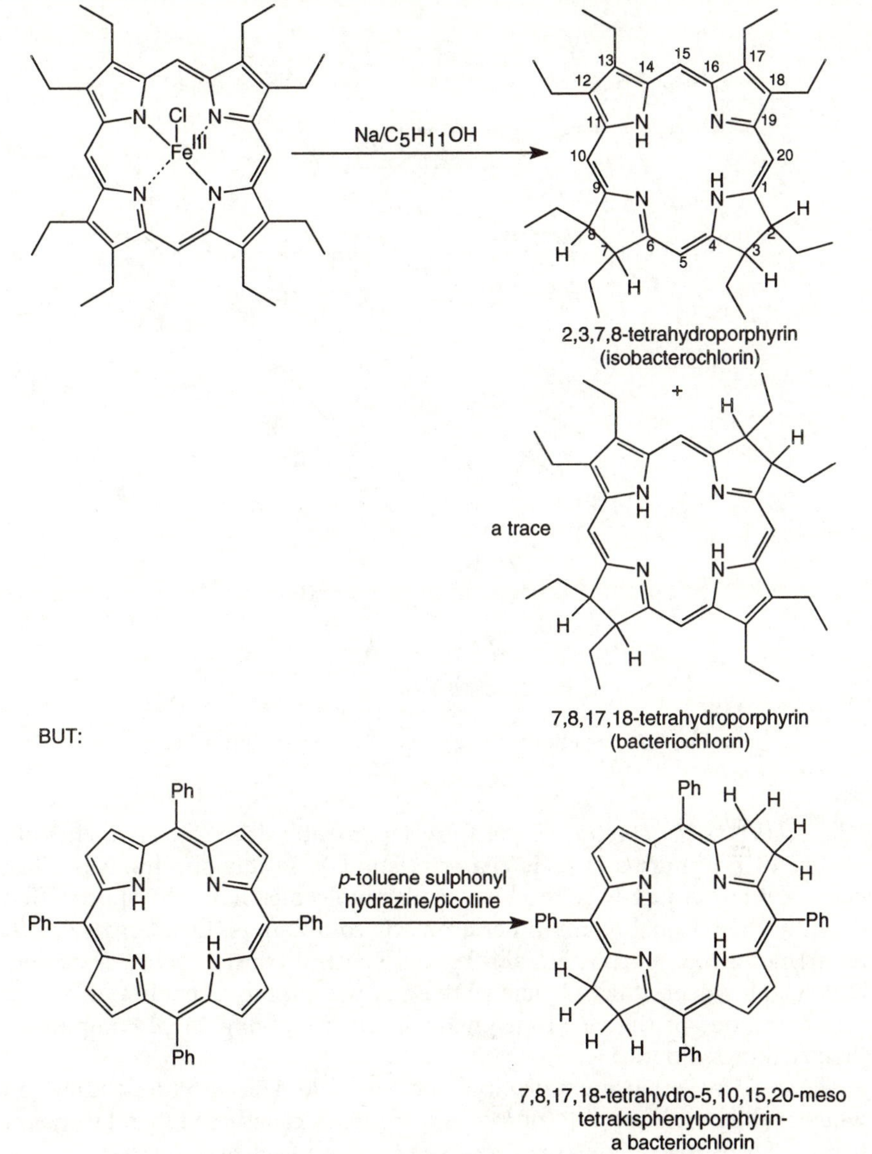

2,3,7,8-tetrahydroporphyrin
(isobacterochlorin)

+

a trace

7,8,17,18-tetrahydroporphyrin
(bacteriochlorin)

BUT:

7,8,17,18-tetrahydro-5,10,15,20-meso
tetrakisphenylporphyrin-
a bacteriochlorin

Fig. 3.21 Reduction of porphyrins to tetrahydroporphyrins.

1,4,6,9,11,14,16, and 19) appear as a broadened peak in the ^{13}C NMR spectra
of porphyrins. However, at $-60\,°C$, two distinct, sharp peaks are observed.
These are assigned to four pyrrole-type α-carbons (e.g. 1,4,11, and 14) and
four pyrrolenine-type α-carbons (e.g. 6,9,16, and 19, see Figure 3.13). In
other words, at $-60\,°C$, the tautomerism is slow on the NMR time-scale.

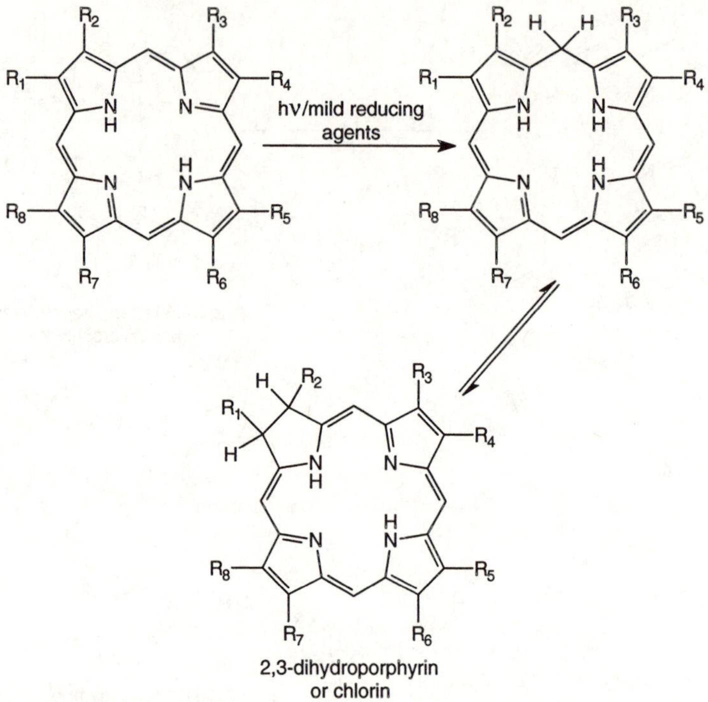

2,3-dihydroporphyrin
or chlorin

Fig. 3.22 Relationship between chlorins and phlorins.

3.1.4.3 UV–visible spectra of porphyrins Probably the most fascinating feature of porphyrins is their characteristic UV–visible spectra. More has been written on just why these particular molecules have the spectra that they do, than arguably any other aromatic compounds. Like benzene, and its homologues, porphyrin absorption spectra consist of two distinct regions. However, unlike benzene, these appear in the near-ultraviolet and visible regions of the electromagnetic spectrum, giving these compounds their striking colours.

There is an intense absorption between 390–425 nm (depending on whether the porphyrin is β- or *meso*-substituted), called the B band (or Soret band, after its discoverer) with between two and four much weaker bands, called Q bands, situated between 480–700 nm. The number and intensity of these bands can give powerful clues to the substitution pattern of the porphyrin and whether it is metalled or not. This has led to several ways of classifying porphyrin spectra.

The pictorial classification of porphyrin spectra relates the number and relative intensity of the Q bands, in the case of an unmetallated porphyrin, to the substituents on the pyrrole β- and *meso*-positions, or, in the case of a metalloporphyrin, to the stability of the central metal cation. Thus, metal-free porphyrins as their free bases have four Q bands, denoted by increasing wavelength as IV, III, II, and I. When the relative intensities of these bands are such that IV > III > II > I, then the spectrum is said to be *etio-type* after the etioporphyrins in which the β-substituents are all alkyl groups [Figure 3.23(a)]. In practice, the etio-type Q band spectrum is found in all porphyrins in which six or more of the β-positions are substituted with groups without π-electrons, e.g. alkyl groups.

Substituents with π-electrons, e.g. electron-withdrawing carbonyl or vinyl groups, attached directly to the β-positions lead to a subtle change in the relative intensities of the Q bands, such that III > IV > II > I. This is called a *rhodo-type* spectrum (after rhodoporphyrin XV) because these groups have a "rhodoflying" or "reddening" effect on the spectrum by shifting it to longer wavelength. Two such groups on pyrrole units next to each other in the macrocycle cancel each other out. However, when these groups are on opposite pyrrole units, then the rhodofying effect is intensified to give an *oxo-rhodo-type* spectrum in which III > II > IV > I. When some of the pyrrole β-positions are left unsubstituted (usually no less than four), or when a *meso*-position is occupied, then the *phyllo-type* spectrum is obtained, in which the intensity ratio of the Q bands is now IV > II > III > I.

Protonation of the porphyrin, with acid, leads to addition of two further protons to the central nitrogens. As far as the macrocycle is concerned, this is a more symmetrical situation than in the porphyrin free base and produces a simplification of the Q band spectrum that can lead to profound colour changes. Essentially, the four Q bands collapse to two.

Porphyrins substituted in the β-pyrrole positions change colour from the wine red of the free base to magenta. However, *meso*-substituted porphyrins go a deep emerald green in solution, coinciding with a red shift in the position of the B band, sometimes by as much as 40 nm, depending on the *meso*-substituent. This has been traced to large conformational changes in the porphyrin macrocycle which result in a tilting of the pyrrole rings about their α-carbons, and a rotation of the *meso*-substituent (usually an aryl moiety) into the plane roughly defined by the α- and meso-carbon atoms.

Metalloporphyrins also have more symmetrical macrocycles than free-base porphyrins, so that their Q band spectra generally consist of only two bands. The pictorial classification denotes these as α- and β bands, the former at longer wavelength than the latter. The relative intensities of these two bands can be a qualitative yardstick of just how stable is the metal complexed to the four porphyrinic nitrogen atoms. Thus, when α > β, the metal forms a stable square-planar complex with the porphyrin, e.g. Ni(II), Pd(II), and Sn(IV). Cadmium(II), on the other hand, which is easily displaced by protons, has β > α.

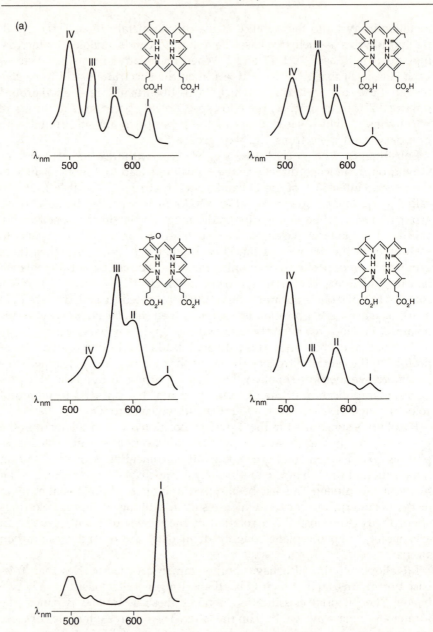

Fig. 3.23 (a) Q band porphyrin spectra for metal-free porphyrins and a chlorin. (b) Q band spectra for porphyrin dications, monocation, and some metal complexes. In the latter, the smaller the ratio between the α and β band intensities the less stable the metal complex.

(b)

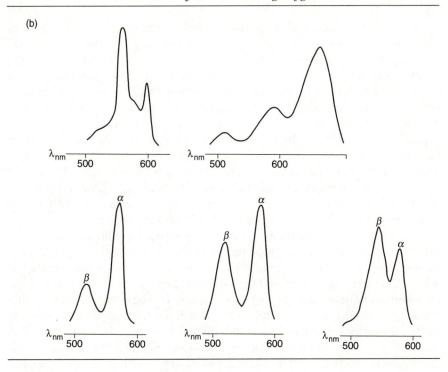

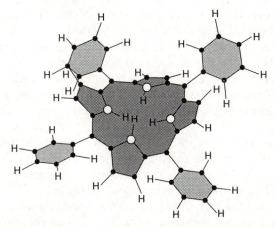

Fig. 3.24 Conformational changes in the porphyrin macrocyclic skeleton (for *meso*-substituted porphyrins only) on dication formation. The pyrrole subunits 'hinge' about carbons 1, 4, 6, 9, 11, 14, 16, and 19 (two up and two down) while the *meso*-aryl units rotate more into the mean macrocyclic plane. [From E. Meyer and D.L. Cullen, in *The Porphyrins*, ed. D. Dolphin, Vol. 3, Academic Press, New York (1978), pp. 513–529.]

The pictorial classification is also useful in comparing spectra of porphyrins and chlorins. Chlorins have one double bond less than porphyrins, and the main spectroscopic difference that results from this lies in the relative intensities of the B and Q bands. The main visible band in chlorins is band I, which is also extensively red shifted (>25 nm to approximately 660 nm) compared with the same band in neutral porphyrins, and is of greatly increased intensity relative to the B band. Thus, in neutral porphyrins and metalloporphyrins, the B band:Q band ratio can be as much as 50:1. In chlorins, however, because of the increased intensity of band I, this ratio may be reduced to about 5:1. This is the reason why chlorophyll absorbs so much more of the visible spectrum of light than porphyrins. After looking more closely at metalloporphin spectra, we shall return briefly to examine how these intensity differences come about.

Metalloporphyrin spectra are classified either as "regular" or "irregular" depending on whether the metal they contain has a closed or open shell of valence electrons.[6] Thus, regular metalloporphyrins give *normal* spectra, i.e. a B band in the near-ultraviolet (390–425 nm depending on the macrocyclic substitution pattern), and two Q bands, α and β. The α band occurs within the range 570–610 nm for complexes in which the macrocycle is substituted in the β-positions. For complexes in which the macrocycle is substituted in the *meso*-positions, the α band occurs between 590 and 630 nm. These positions depend on the metal that is complexed and whether that metal carries any axial ligands.

Regular metalloporphyrin complexes fluoresce, with their fluorescence quantum yield modified by the heavy-atom effect. The complexes tend to be purple in the solid state and wine red in solution. The *normal* absorption spectra of regular porphyrins is indicative of little or no interaction between atomic orbitals on the metal and π-molecular orbitals on the porphyrin. Thus, the absorption and emission spectra of regular metalloporphyrins are largely determined by the porphyrin's π-electrons. This is not the case with irregular metalloporphyrins. They show three main types of spectra called *normal*, *hypso*, and *hyper*.

Normal-type spectra are shown by metals of the d- or f-block where the metal d- or f-electrons are of such low energy that they do not interact significantly with the porphyrin π-electrons; for example, the vanadyl cation, VO^{2+} (d^1) and the europium cation, Eu(III) (f^6).

Hypso-type spectra look similar to normal spectra but the B and Q bands are blue shifted (i.e. hypsochromically). This type of spectrum is shown by d-block elements with unfilled d-orbitals of the type d^6–d^9, which includes all the metals from groups VIII to IB. Here, d-electrons may be donated into the porphyrin's empty π*-orbitals (i.e. metal-to-ring charge transfer), thus raising their energy. This increases the energy of the porphyrin π–π* transition, with respect to the metal-free porphyrin, leading to a blue (hypsochromic) shift of the spectrum. This effect increases with increasing atomic number of the transition metal, e.g. in the series Ni(II), Pd(II), and Pt(II).

Hyper-type spectra are abnormal in that other intense absorptions in the ultraviolet region are present in addition to the B and Q bands. These spectra come in two types—*p-type* and *d-type* hyper spectra—depending on whether the central cation is a main group element from groups IV or V in an oxidation state two below the group number [i.e. with an open shell outer electronic configuration, e.g. Sn(II), Pb(II), P(III), As(III), Sb(III), and Bi(III)], or a transition metal in which the number of d electrons is no greater than five [e.g. Fe(III), Mn(III), Cr(III), Mo(V), and W(V)]. The p-type hyper spectra arise by charge transfer from the metal p-orbitals into the empty porphyrins π^*-orbitals. The extra bands found in d-type hyper spectra have a different origin. Here, the charge transfer is from the filled porphyrin π-orbitals into vacancies in the transition metal's d-orbitals (i.e. ring-to-metal charge transfer). Figure 3.25 shows a typical hyper-type spectrum for a Mn(III) porphyrin.

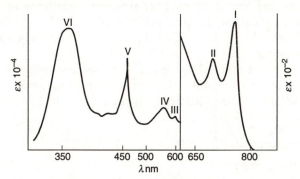

Fig. 3.25 Typical d-type hypo-type UV–visible spectrum of an Mn(III) porphyrin.

3.1.4.4 *Origins of metal-free spectra*

Owing to the central importance of porphyrins and related compounds in biology, there have been many attempts over the years to derive from theory the main facts of porphyrin UV–visible spectra, i.e. the positions and multiplicities of the B and Q bands. Foremost in this search has been the American theoretical chemist, Martin Gouterman. He developed a theory of metal-free porphyrin spectra that allows an intuitive appreciation of the light-induced electronic movements that occur inside porphyrins and related molecules. Ultimately, this begins to answer questions such as why chlorins are used as sensitisers of photosynthesis.

Gouterman's theory is called the four-orbital model,[7] because it considers only the two highest occupied (with electrons) molecular orbitals (HOMOs) of a porphyrin, and the two lowest unoccupied molecular orbitals

(LUMOs). Other, more sophisticated models have now superseded it (in particular, models that can include the effects on the spectra of complexed metal ions, and vibrations within the macrocycle), but the four-orbital approach still serves as the best introduction to the theoretical complexities of porphyrin UV–visible spectra. First, and to put the four-orbital model in context, it will be interesting to look at some of its antecedents.

Free electron theory[6] (FET) treats the excited electron as if it were a particle rotating around a ring. In this case, the ring contains 18 lattice points (atoms), and the excited electron goes round this ring clockwise or anti-clockwise. In so doing (and remembering that the motion of the electron is quantised), it generates rotational electronic energy states which, except for the ground state, are all doubly degenerate with increasing orbital angular momentum.[8] The orbitals generated are shown in Figure 3.26. The ground state has angular momentum, $L = 0$. By analogy with benzene, electrons are placed in orbitals of increasing angular momentum, two in $L = 0$ and four in each of the $L = \pm1, \pm2, \pm3, \pm4$. After a mathematical process called configuration interaction (CI; we shall look at this later), the lowest energy excited states are formed from four possible transitions of one electron between the two highest occupied orbitals (i.e. $L = \pm4$), and the two lowest unoccupied orbitals (i.e. $L = \pm5$). This causes a change in angular momentum of the electron of $\Delta L = \pm1$ or ±9.

Now, according to the orbital angular momentum selection rules (which determine what transitions the electron is allowed to make), the only electronic transitions that are allowed are those in which either the orbital

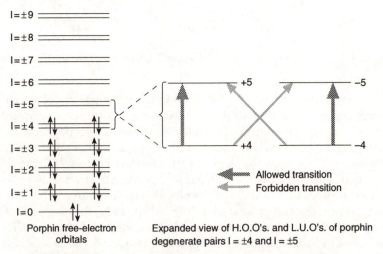

Porphin free-electron orbitals

Expanded view of H.O.O's. and L.U.O's. of porphin degenerate pairs l = ±4 and l = ±5

Fig. 3.26 & 27 Orbitals generated by free electron theory. Allowed and forbidden transitions in free electron theory.

angular momentum does not change, or, if it changes, it only does so by one unit. Mathematics puts this statement far more succinctly: $\Delta L = \pm 1$ or 0. This means that the $\Delta L = \pm 9$ transitions are forbidden, according to the selection rules. So, there are now four transitions of equal energy; two are allowed (± 1) and two are forbidden (± 9). After CI is applied, one degenerate pair of states (the allowed pair, with $L = \pm 1$) is now at higher energy and accounts for the B band, while the other degenerate pair of states (the forbidden pair, with $L = \pm 9$) are at lower energy and accounts for the Q bands (Figure 3.27).

The FET applied to porphyrins, therefore, predicts that the electronic transitions that account for the B and Q bands occur in the right region of the spectrum. It also correctly predicts the relative intensities of these bands. That is to say, the more allowed an electronic transition is, the more intense will be the absorption band it produces. So, the allowed electronic transitions give rise to the intense B band, while the forbidden electronic transitions account for the weak Q bands. However, the FET cannot predict the multiplicity (i.e. the number) of Q bands. In order to do this, it had to introduce the idea that a porphyrin exists as two tautomeric forms: one in which the central hydrogens are on opposite nitrogen atoms, and the other in which these hydrogens reside on adjacent nitrogen atoms (Figure 3.28).

X-ray crystallography put paid to that idea, showing that the central

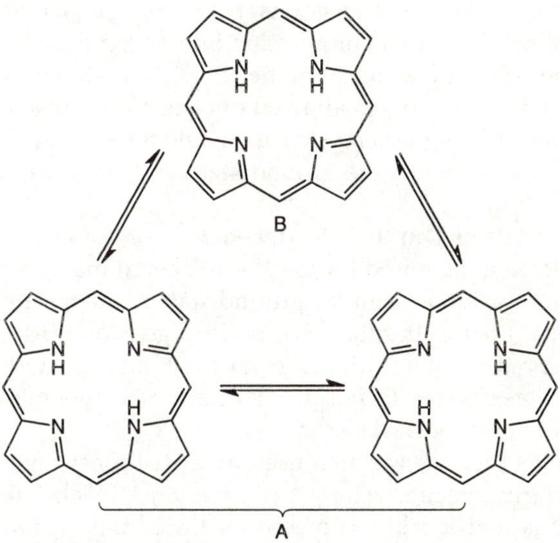

Fig. 3.28 In order for free electron theory to be able to predict the multiplicity of Q bands, it had to suggest that the macrocycle exists in two interchangeable tautomeric forms, A and B.

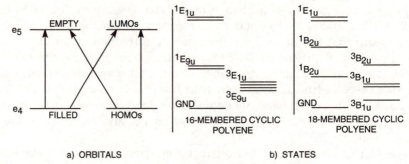

Fig. 3.29 Lowest energy states for transitions between HOMOs and LUMOs of 16- and 18-membered cyclic polyenes with 18 π-electrons.

hydrogen atoms exist on opposite hydrogen atoms only. If the hydrogen atoms were adjacent to each other, they would sterically hinder each other. In other words, the central hole of the porphyrin is simply not big enough to accommodate two adjacent hydrogen atoms. Another thing that the FET cannot do is to predict how changes to the porphyrin skeleton, e.g. in chlorins and phthalocyanines, affect the spectra.

The cyclic polyene theory[6] (CPT) treats the porphyrin macrocycle as either a 16-membered (metal complex or dication) or an 18-membered (freebase) cyclic polyene. In this treatment, a set of molecular orbitals are generated very similar to those of the FET, in which degenerate excited states are generated by one-electron transitions between degenerate HOMOs and LUMOs. Application of the configuration interaction lifts the degeneracy of the excited states, giving a pair of higher energy and a pair of lower energy states. The B band arises by an allowed one-electron transition between the ground state and the higher energy pair, while the Q bands come from forbidden transitions between the ground state and the lower energy pair of excited states.[9]

CPT differs from FET in that the degeneracy of the lower energy pair of states is predicted to be raised for the 18-membered macrocycle. This means that forbidden transitions from the ground state to these states will generate more than one Q band. In other words, CPT goes one better than FET, in being able to predict the multiplicities of the B and Q bands. However, as with FET, CPT as applied to porphyrins cannot be generalised to consider variations in the porphyrin skeleton.

Simple Hückel theory[10] was first used by Christopher Longuet-Higgins to predict porphyrin spectra.[6] The great advantage Hückel theory has over FET and CPT is that it takes into account the actual geometry of the porphyrin macrocycle. In using this approach to calculate the HOMOs and LUMOs between which one-electron transitions take place, Longuet-Higgins obtained two non-degenerate HOMOs (labelled a_{2u} and a_{1u} for

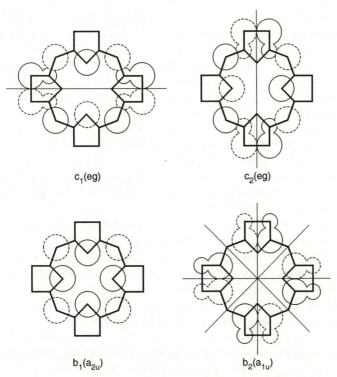

$c_1(eg)$ $c_2(eg)$

$b_1(a_{2u})$ $b_2(a_{1u})$

Fig. 3.30 Porphyrin HOMOs b_1, b_2 and LUMOs c_1 and c_2 showing their original (Hückel) group theoretical notation and their modified notation as used on the four-orbital model. The original coefficients are proportional to the size of the circles. Solid or dashed circles indicate sign. Symmetry nodes are drawn in heavy lines.

symmetry reasons) and a degenerate pair of LUMOs (both labelled e_g). Calculations put the a_{2u}–e_g transition at lower energy than a_{1u}–e_g, so that the former was identified with the Q bands, and the latter with the B band. However, these same calculations showed that both transitions had roughly similar absorption strengths, contrary to what is actually observed. Although this is clearly wrong, the advantage of the Hückel method was that it did allow (albeit on a qualitative basis only) some predictions to be made of how variations in the porphyrin skeleton, e.g. ring reduction in the chlorin series, affect optical transitions.

So, to summarise, FET and CPT produced surprisingly accurate but qualitative accounts of porphyrin spectra which could not, however, be generalised to consider variations in the porphyrin skeleton. On the other hand, simple Hückel theory, which took account of the detailed shape of the porphyrin macrocycle, wrongly predicted the relative intensities of the

B and Q bands, but did provide qualitative data on how skeletal variations affect porphyrin absorption spectra. The stage was now set for the strengths of all these theories to be unified. Martin Gouterman did this by using configuration interaction, a technique vital in free electron and cyclic polyene theories, to describe the porphyrin excited states generated by Hückel theory. He called this new synthesis the four-orbital model.[6,7]

The four-orbital model uses the two HOMOs and LUMOs generated by simple Hückel theory. Gouterman relabelled them so that the a_{1u} orbital becomes b_2, while a_{2u} becomes b_1. This was done from symmetry considerations: it allows the same orbitals to be used in situations where there is a reduction in the symmetry of the porphyrin macrocycle. This happens, for example, in going from the metal complex or dication (which belong to the D_{4h} point group[11]), to the porphyrin free base (which is of lower symmetry and belongs to the D_{2h} point group). At this point, inspection of these Hückel molecular orbitals shows that they have something in common with those generated by cyclic polyene theory. The b_1 and b_2 orbitals have four nodes, while the c_1 and c_2 orbitals have five. This means that the Hückel orbitals and the cyclic polyene orbitals are topologically the same, just like a circle and an ellipse. Now comes Gouterman's first big assumption. Why not make the Hückel HOMOs accidentally degenerate? This means that the Hückel orbitals are now even more like those of the cyclic polyene; they also have the same energy. It also means that the configuration interaction can be used to generate the excited states corresponding to the main porphyrin absorption bands.

Before going into that, let us look more closely at the configuration interaction. This is a mathematical technique that allows the best (i.e. lowest energy) orbital wave functions for the ground and excited states to be found. It is done by mixing various bonding and antibonding wave functions having the same symmetry. The result is a set of new wave functions, the most stable of which is more stable than any individual contributing structure. Thus, a good approximation to the ground state configuration will be a structure where the electrons are in the lowest energy molecular orbital. The contribution made by higher energy excited states to any particular structure turns out to be inversely proportional to their energy relative to the ground state.

So, one-electron transitions between Gouterman's two new HOMOs (now accidentally degenerate) and two LUMOs, generates four degenerate electronically excited states. Configuration interaction then mixes them to give a sum and a difference of intensities which were originally equal.

At this stage, it is necessary to introduce another concept: the transition dipole. It is directly related to the strength of light absorption by a molecule; the greater the transition dipole, the greater the absorption. The transition dipole arises because when an electron is excited, charge separa-

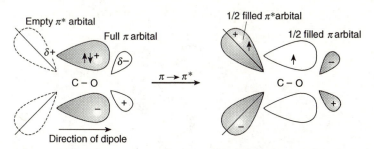

Fig. 3.31 In a carbonyl group the π-orbital is distorted towards the more electronegative oxygen atom. This gives the group a negative dipole towards the oxygen atom. When a π–π* transition takes place, an electron enters the π*-orbital which has more electron density over the carbon atom. Therefore, during excitation (or relaxation) the dipole changes.

tion and redistribution occurs. This can easily be seen by considering what happens when a carbonyl group absorbs energy, causing an electron to be excited out of a π-orbital into a π*-orbital.

In the π-orbital, because of the higher electronegativity of the oxygen atom, the electron spends more time near it. Consequently, the molecule exhibits an electric dipole along the C=O axis. However, when electronic excitation takes place, the π*-orbital is occupied and now the excited electron spends more time near the carbon atom. Therefore, in the excited state, the polarity of the C=O bond is reduced. In undergoing the electronic transition, the C=O bond experiences a strong transition dipole which can be resolved into two components whose directions are perpendicular to each other. Now, back to the four orbital model.

If b_1c_1 represents the singlet state produced by the transition b_1 to c_1, this will have a similar symmetry to the state b_2c_2. Also, the singlet state b_1c_2 has similar symmetry to b_2c_1. The sum and difference of intensities will be represented by:

$$\tfrac{1}{2}(b_1c_1 + b_2c_2) = B_y^{\,0}; \tfrac{1}{2}(b_1c_2 + b_2c_1) = B_x^{\,0}$$
$$\tfrac{1}{2}(b_1c_1 - b_2c_2) = Q_y^{\,0}; \tfrac{1}{2}(b_1c_2 - b_2c_1) = Q_x^{\,0}$$

where x and y represent the directions of the components of the transition dipoles produces as the electron is excited from ground to excited states. For a porphyrin dication or a metal complex, the x and y directions are equivalent, but for a free-base porphyrin the x direction is defined along the NH–NH axis: the y direction is perpendicular to this. B_y and B_x represent the sum of the intensities of the electronic transitions and correspond to a strongly allowed transition giving rise to the B band. Similarly, Q_y and Q_x represent the difference of the intensities of the electronic transitions. Now,

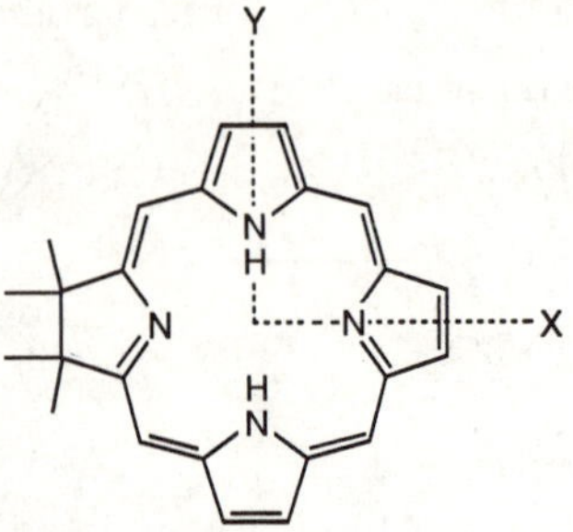

Fig. 3.32 Chlorin transition dipole directions.

one of Gouterman's basic assumptions is that the HOMOs b_1 and b_2, as well as the LUMOs, are degenerate. Consequently,

$$b_1c_1 - b_2c_2 = b_1c_2 - b_2c_1 = 0$$

In other words, the Q bands are forbidden and simply shouldn't exist. The fact that they do is because of molecular vibrations within the porphyrin macrocycle. These have the effect of marginally lifting the degeneracy of b_1 and b_2 so that the difference in intensities represented by Q_y and Q_x is no longer equal to zero, i.e. the Q bands become weakly allowed. Their weakness allows them to show vibrational fine structure. In the case of a porphyrin dication or metal complex, where the x and y directions for the components of the transition dipoles are equivalent, this gives rise to two Q bands. For porphyrin free bases, where the x and y directions are perpendicular to each other, each component has two associated Q bands, so the total number is four.

The four-orbital model, therefore, accounts for the number, multiplicity, and relative intensity of the Q bands. It also accounts correctly for how the intensity of the Q bands varies when the porphyrin complexes with a metal cation or picks up two further protons to become a dication.

Inspection of the porphyrin HOMOs and LUMOs show that only the b_2 orbital has no electron density over the central nitrogens. Consequently, this will be the only orbital to be energetically unaffected by metal chelation or dication formation. Electrons on the metal will electrostatically repel the electrons in the remaining orbitals and so raise their energies. Conversely, the two positive charges on the central nitrogens (which arise on dication formation, from the addition of two extra protons) will electrostatically attract the electrons in the b_1 and b_2 orbitals and so lower their energy.

The overall effect is the same in both cases. The degeneracy of c_1 and c_2 will be unaffected but the degeneracy of the HOMOs b_1 and b_2 will be lifted by an amount that depends on the strength of the interaction of the metal or protons with the central nitrogens. Again, the lifting of the degeneracy of

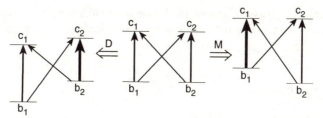

Fig. 3.33 Lifting of degeneracy of porphyrin HOMOs on (M) metal chelation and (D) dication formation: b_1, b_2, c_1, and c_2 represent the MOs in Figure 3.30. On metal chelation (M), b_1, c_1 and c_2 are raised in energy. On dication formation (D), b_1, c_1, and c_2 are lowered in energy.

these orbitals (more serious than in the case of molecular vibrations) will lead to the Q bands becoming more allowed and so increasing in intensity. This interaction is quite large in *meso*-substituted porphyrin dications so that the relative intensity of Q bands to B band is now about 1:5 instead of the usual 1:25–50 for porphyrin free bases. In the limiting case, where the degeneracy of b_1 and b_2 is severely lifted, the B and Q bands are of nearly equal intensity, realising the simple Hückel theory prediction.

Another interesting case for the application of the four-orbital model is interpretation of chlorin spectra. Gouterman here chooses ring IV to be reduced so that pyrrole positions 17 and 18 are removed from the conjugation pathway. By considering all the relevant resonance structures, Gouterman showed that this forces the N–H axis to be parallel with the y axis of the transition dipole component. Now the orbitals most affected by reductio of ring IV, will be those with electron density over positions 17 and 18 in the fully conjugated macrocycle. These are b_2 and c_1. They will be raised in energy, compared with b_1 and c_2, because reduction at positions 17 and 18 reduces the size of their "free electron box". In other words, reduction causes electrons in these orbitals to be more confined, which raises the orbital energy.

Electronic transitions polarised along the x-axis (i.e. b_1 to c_2 and b_2 to c_1, see Figure 3.33) turn out to be nearly degenerate, so that their sum and difference produce, respectively, a strong B_x band and a weak Q_x band. In the y-polarised direction, however, there is a great difference in energy between the respective electronic transitions (i.e. b_1 to c_1 and b_2 to c_2) so that their sum and difference produce, respectively, B_y and Q_y bands of comparable energy. The B_x and B_y bands coincide in energy and intensity (as they do in the porphyrin series), but now the Q_y band is of much greater intensity than the Q_x band.

Thus, the four-orbital model predicts an intense band at long wavelength, and is so able to explain the green colour of chlorins. The heart of chlorophyll is a metallochlorin macrocycle, so that we are now in a position to understand why grass is green. In the next section we shall consider how chlorophyll achieves the first steps of photosynthesis, i.e. trapping of light energy, and how oxygen is produced.

3.2 Making oxygen

3.2.1 Introduction

Without oxygen, life could never have progressed as far as it has from its primeval antecedents. Yet it can come as quite a shock to realise that oxygen is, in effect, a by-product of a life process. That process is photosynthesis, i.e. the conversion of carbon dioxide and water into carbohydrates and oxygen, through the intervention of light.

The point here is that carbon dioxide and water are low energy molecules, whereas carbohydrates are high-energy molecules. This reverses the natural tendency of high energy molecules to react to form low energy species. In other words, photosynthesis works against the prevailing thermodynamic gradient. Without photosynthesis, the world would be a lifeless and airless place in which all the free oxygen would have reacted eons ago, to become locked up in rocks, as oxides and silicates, or forming a poisonous, heat-trapping atmosphere consisting of the gaseous oxides of carbon, sulphur, and nitrogen—a scenario currently being played out on a planet near us called Venus. Life, as we know it, results from a waste product of photosynthesis, oxygen. How?

3.2.2 The photosynthetic apparatus

This apparatus is contained within the leaf or the blade of grass. Look at some pond water under a microscope and you will see small, green, single-celled, plant-like animals: they are green and perform photosynthesis like plants, but they move around like animals, e.g. *Euglena* and *Volvox*. They too contain the photosynthetic apparatus. So what is it and where is it located?

Under a light or electron microscope, cells of higher plants and the green algae contain small subcellular structures called *chloroplasts*. These contain the chlorophyll pigment molecules, which absorb light energy, and all the enzymatic and chemical reactions that use this energy to convert carbon dioxide and water into carbohydrates.

Chloroplasts are approximately oval saucers, 4–10 μm in diameter and about 1 μm thick.[12] Each has a double membrane (the chloroplast envelope) around it which separates it from the rest of the plant cell. Inside, the chloroplast contains stacks (known as *grana*) of green discs, called *thylakoids*, that look like miniature blocks of flats joined together by "walk-

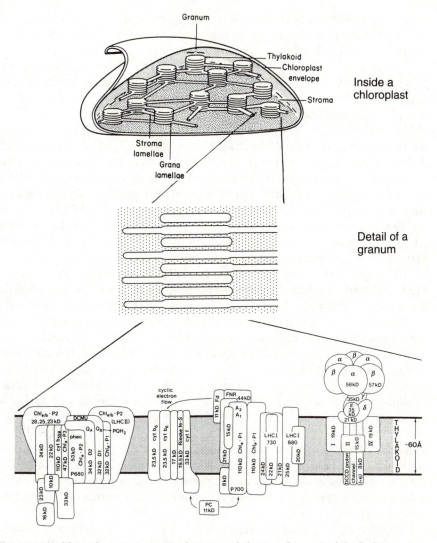

Fig. 3.34 A chloroplast, its inner workings, and the 'machinery' of the light reaction in the thylakoid membrane. [From D.O. Hall and K.F. Rao, *Photosynthesis*, 4th Edn Edward Arnold, London (1988) p. 98.]

ways" or *stroma lamellae*. Each thylakoid may contain two double layer membranes which contain the photosynthetic pigments in the form of protein complexes. The rest of the chloroplast is a colourless region called the *stroma* where, in conjunction with the necessary enzymes, carbohydrate production is situated.

3.2.2.1 Photosynthesis as two reactions   The internal division of chloroplasts into two distinct regions (the grana, containing the photosynthetic pigments and the colourless carbohydrate-producing stroma) suggests that photosynthesis might also consist of two reactions. This was found to be the case as early as 1905, when it was discovered that photosynthesis is a two-step mechanism consisting of a photochemical, or *light reaction*, and a non-photochemical or so-called *dark reaction*.

The light reaction consists of (a) trapping of light energy by chlorophyll molecules, (b) conversion of that trapped energy into a strong reducing agent (NADPH) and the generation of ATP, and (c) concomitant oxidation of water to oxygen as a by-product. In the dark reaction, enzymes utilise the NADPH and the ATP generated during the light reaction to produce carbohydrates. It should be emphasised that as long as there is a supply of NADPH and ATP the dark reaction will proceed, be it pitch black or broad daylight. The light reaction requires a source of illumination to work at all. It just so happens that the end products of the light reaction are those two useful reagents, NADPH and ATP. Eons ago nature tried the experiment of coupling the light and dark reactions together. The rest, as they say, is exceptionally ancient history. It was also the first time that a living process had emitted a toxic pollutant into the (then) biosphere. That pollutant was oxygen.[13]

Imagine that a new strain of algae suddenly appeared; one that, for arguments sake, converted chloride ions in the environment into chlorine. Also imagine that the biochemistry of this truly terrifying (but thankfully hypothetical) organism was somehow able to withstand the ravages of chlorine. Obviously, those organisms that couldn't cope with chlorine (i.e. all life as we know it) would have a death sentence over them as the algae bloomed and the levels of atmospheric and aquatic chlorine built up (for good measure, chlorine dissolves in water generating hydrochloric and hypochlorous acids; the latter being more commonly known as bleach). Allegorically, that was the situation facing early life forms about 1–2 billion years ago, when photosynthetic organisms were first thought to have appeared that could oxidise water to oxygen.

The very first photosynthesisers did not produce oxygen and utilised electron donors other than water. They existed in a reducing atmosphere probably similar to that known to exist on the big gas planets like Jupiter and Saturn, and they would have thrived on a plentiful supply of organic compounds and carbon dioxide from anaerobic fermentation processes going on in the primeval soup. Such organisms still exist and they contain only one type of chlorophyll, called chlorophyll a_{683} (because they have an absorption maximum at 683 nm). Their chlorophyll is organised into a photosystem (called photosystem 1 or PS1) which is a collection of protein particles that contains all the photosynthetic pigments embedded in the cell membrane.

The photosystem consists of a light-harvesting unit of about 200 chlorophyll a_{683} molecules and approximately 50 β-carotene molecules arranged in an array whose function is like a molecular Jodrell Bank. They collect and direct the harvested light to a specialised pair of chlorophyll molecules, called chlorophyll P700 (their absorption maximum is at 700 nm), which traps the collected light energy.

Photosynthetic organisms that oxidise water, however, have *two* photosystems. They have PS1 but also another, called, not surprisingly, PS2, which actually has two types of chlorophyll, chlorophyll a_{673} (which is structurally the same as a_{683}) and chlorophyll b, the latter having one group different from chlorophyll a. Together, the two chlorophylls enable the photosystem to absorb light energy over a larger spread of wavelengths.

The two photosystems, each a collection of protein particles containing the photosynthetic pigments, are embedded in and separated by the thylakoid membrane. Connecting to and associated with each of these photosynthetic units are chains of electron-transporting molecules consisting of *quinones* and heme-containing proteins called *cytochromes*.[14] This shows that the light reaction is itself two coupled photochemical reactions, performed by PS1 and PS2. Without PS2, the photosynthetic oxidation of water to oxygen cannot take place.

When this reaction first appeared on the planet, some 1–2 billion years ago, it spelt ultimate doom for the then major life form, anaerobic bacteria. It is thought that the oxygen levels did not build up gradually. The reducing atmosphere of the time ensured that iron compounds existed in their more water soluble Fe(II) form. Thus, the seas would have contained a high concentration of Fe(II) compounds. As the aerobic organisms multiplied, the oxygen they produced was absorbed by converting water-soluble Fe(II) into red, insoluble Fe(III). Thus, oxygen levels would have remained low until all the Fe(II) was titrated out of the oceans. Then, oxygen levels would have shot up to near-present day levels, giving the anaerobes no time to adjust. They either underwent massive extinction, retreated to parts of the planet where oxygen couldn't reach, or joined forces with the new aerobic organisms. There is evidence that some of the specialised organelles within the cells of aerobic organisms (e.g. mitochondira, the cellular power station where the energy of the cell is generated) may be the descendants of anaerobes that went into partnership with the new photosynthesisers. And it was well that they did, for it was not until the appearance of organisms with PS2 activity, and concomitant oxygen production from water photolysis, that the evolution of complex life forms could really get into its stride. By the beginning of the Cambrian era, some 600 million years ago, the first hard-shelled organisms resembling arthropods had appeared.

Interestingly, the biosynthesis of chlorophyll may be a window looking back into evolution.[15] In this biosynthetic pathway, porphyrins are formed

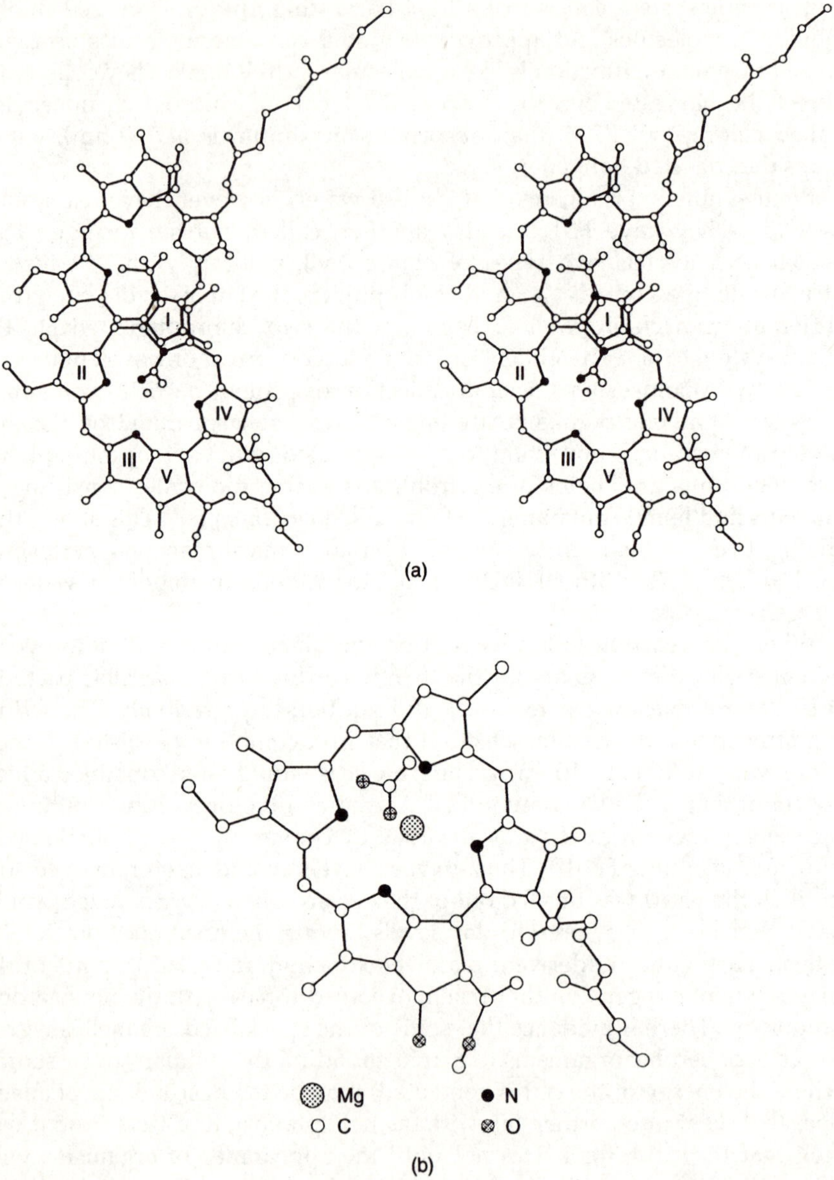

(a)

◉ Mg   ● N
○ C   ⊗ O

(b)

Fig. 3.35 (a) Stereoview of the special pair in the photoreaction centre. Rings I of the chlorophyll molecules are stacked upon each other, and the magnesium atom of each chlorophyll is coordinated by an acetate group from the other molecule. (b) Close up of the nearer chlorophyll molecule from part (a). The unattached acetate group is from the other chlorophyll molecule. (Taken from Huheey *et al.*)

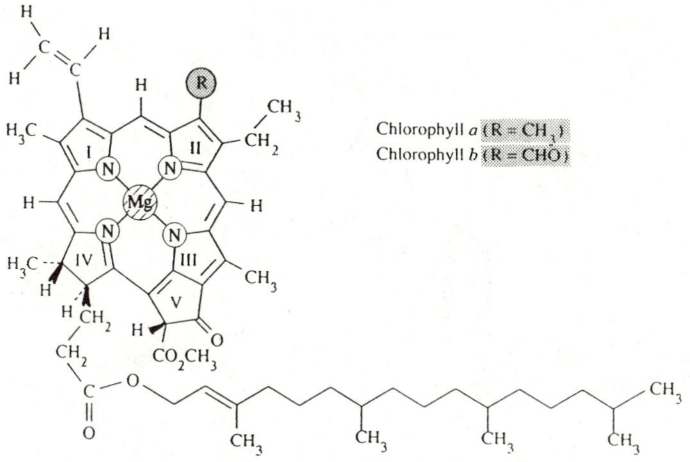

Fig. 3.36 Chlorophylls *a* and *b*.

first, then metalloporphyrins, and finally chlorophyll. Now, in the absence of oxygen, the favoured photochemical reaction of porphyrins (which work well in the UV light that would have bathed the earth's surface before oxygen, and hence a UV-blocking ozone layer, appeared) is to oxidise organic compounds. Metalloporphyrins and chlorophylls reduce them. Thus, primitive photosynthesis probably utilised porphyrins, while modern photosynthesis uses chlorophyll. Moreover, the direction of evolution has been to form highly organised and efficient cellular structures associated with hydrophobic proteins and lipids bound inside double-layer membranes. Life most probably originated in the sea, however, so that a requirement of an early photosynthetic porphyrin would be high water solubility. Like some ancestral echo, the first porphyrin formed in the biosynthesis of chlorophyll is the highly water-soluble uroporphyrin.

3.2.2.2 The light reaction as a "two-stage pump" Photosystems 1 and 2 are embedded in the thylakoid membrane and linked together by a chain of electron-transporting proteins. They act together, therefore, as a highly efficient light-activated, two-stage electron pump which, in effect, boosts the energy of electrons through about 1.2 V. This is ultimately used to generate the strong reducing agent, NADPH. At the same time, the "pump" generates ATP via side reactions coupled to the electron-transfer process between the photosystems. These two reagents are the chemical tools that convert carbon dioxide into carbohydrates in the dark reaction, and as such are the initial products of the conversion of light energy into chemical energy.

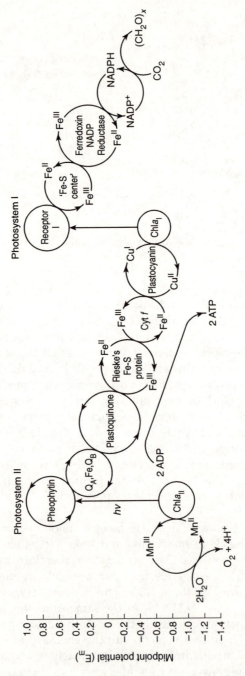

Fig. 3.37 PS1 and PS2. [Huheey *et al.*]

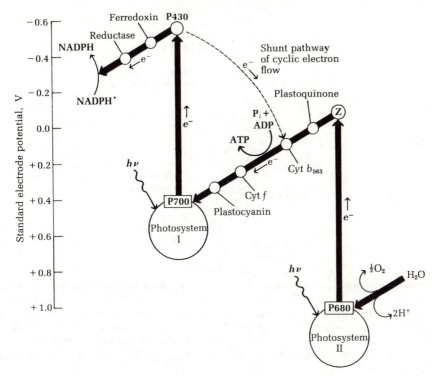

Fig. 3.38 Light reaction as a two-stage electron pump.

3.2.2.3 Antennae and traps Initially, photons of light energy are harvested by the massed chlorophyll molecules of the PS2 antenna system, and fed to the chlorophyll P680 trap, which becomes photoexcited. This happens because the photosynthetic pigments of PS1 and PS2 are arranged into highly efficient light-harvesting and energy-transferring mechanisms. If one of the pigment molecules absorbs a photon, then the excitation energy is rapidly (i.e. about 10^{-15}s) transferred to the rest of the array of pigment molecules. Two transfer mechanisms are possible which differ in their dependence on the separation between pigment chromophore molecules.

The first mechanism, called *exciton delocalisation*, occurs when the inter-chromophore separation is about 1–2 nm, and is dependent on the inverse cube (i.e. r^{-3}) of that separation. What happens is that when the chromophore molecules are held rigidly together (as they would be, more or less, in the chloroplast membrane) at about 1 – 2 nm, their excited states become perturbed enough to form a new set of excited states that are delocalised over the whole array of chromophores.[16] Thus, as soon as an electron is excited, it is delocalised over all the pigment molecules (Figure 3.39).

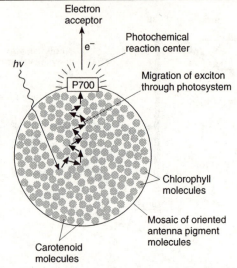

Fig. 3.39 Mechanism of excitation delocalisation.
Schematic diagram of the surface of a photosystem in the thylakoid membrane. It contains a patch-like mosaic of several hundred chlorophyll and carotenoid antenna molecules oriented in the membrane. An exciton absorbed by an antenna molecule quickly migrates via the pigment molecules to the reaction centre, P700; its path is shown by the coloured arrows. Although all the antenna molecules can absorb light, only the reaction centre molecule can convert the excitation energy into electron flow.

The second mechanism, called *Förster* or *resonance transfer*, occurs when the interchromophore separation is about 5 – 10 nm, and is dependent on the inverse sixth (i.e. r^{-6}) of that separation. The rearrangement of electrons that occurs when a photon is absorbed by a porphyrin molecule and it is electronically excited, gives rise to a transition dipole (as noted earlier). When an excited state of a porphyrin molecule decays, the electron rearrangement that occurs gives rise to a transition dipole in the reverse direction. This, in turn, can induce a transition dipole in a neighbouring molecule up to several molecular diameters away.[17a] This dipole–dipole interaction will happen if the relevant electronic transitions are allowed and if there is some overlap between the fluorescence spectrum of one chlorophyll molecule and the absorption spectrum of the other. In other words, when the energy gaps between ground and excited states of the two interacting molecules are comparable (Figure 3.40). The trapping centres, P700 for PS1 and P680 for PS2, have energy gaps between their ground and excited states that are less than those of the rest of the pigment array. Consequently, the excitation that is delocalised over the whole array, is trapped here.

Some support for these ideas comes from recent results about the relative placement of chromophores in a bacterial photosynthetic reaction centre.[17b]

Fig. 3.40 Mechanism of Förster transfer. Excitation of one chlorophyll molecule induces transition dipoles (and hence excitation in nearby molecules).

Here, two bacteriochlorophyll (BCl) molecules are closely associated both spatially and electronically, to form the primary "special pair" donor, P865. Another pair of BCl molecules are placed edge to edge, relative to P865, and these in turn are adjacent to a pair of bacteriopheophorbide a (BPh) molecules. A pair of quinones next to the BPh molecules completes the picture (Figure 3.41).

Light excitation of any of the macrocyclic chromophores within the reaction centre protein results in energy transfer to P865 in less than 150 fs. Once P865 is excited, then electron transfer to the BPheo a to the left of P865 occurs in 2.8 ps. Why the BPheo a molecule on the right is not used is not yet clear. Electron transfer to a quinone then occurs in about 200 ps, so that in only two or three sequential electron-transfer steps, the separated charge has crossed about 3.5 nm to give an intermediate with a lifetime of about 100 ms.

In green-plant photosynthesis, the excitation of an electron from the trapping chlorophyll P680's ground state to an excited state boosts its energy by about 0.8 V. The "positive hole" that is left in this trapping species ground state can accept an electron from a suitable electron donor, which as we shall see is ultimately water. Meanwhile, the excited electron is donated to an electron acceptor called *plastoquinone*. Thus, the excited state of chlorophyll P680 behaves as an efficient electron donor and acceptor. The overall effect at this stage of the "pumping" cycle is to boost the energy of the electron by about 0.8 volts and to generate a strong oxidant (the oxidised P680) and a weak reductant (the reduced plastoquinone).

Plastoquinone is the beginning of the electron-transport chain that connects PS1 to PS2. From plastoquinone, the electron is transported, via a sequence of hemoproteins, e.g. cytochromes b_6 and f, and a blue copper-containing protein called *plastocyanine*. This is the last link in the electron-transport chain which ends on the reaction centre of PS1, chlorophyll P700. However, in travelling along the electron-transport chain, the electron has, energetically speaking, fallen downhill by about 0.4 V. The energy liberated by the fall is used to do useful biochemical work, by converting ADP into ATP (between cytochromes b_6 and f) in a process known as *non-cyclic phosphorylation*.

Excitation of chlorophyll P700 further boosts the energy of the electron by about another 0.8 V. This creates a positive hole in the ground state of P700

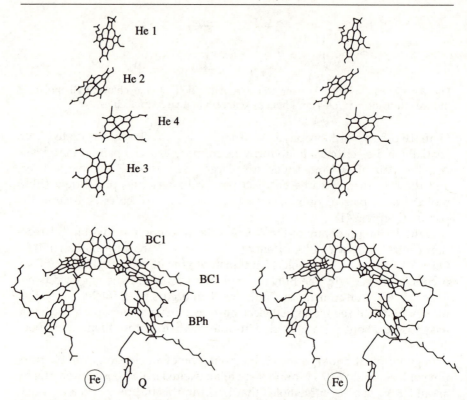

Fig. 3.41 Stereoview of the photosynthetic reaction centre. The photoexcited electron is transferred from the special pair to another molecule of bacteriochlorophyll (BC1), then to a molecule of bacteriopheophytin (BPh), then to a bound quinone (Q), all in a period of 250 ps. From the quinone it passes through the non-heme iron [FeO to an unbound quinone (not shown)] in a period of about 100μs. The electron is restored to the 'hole' in the special pair via the chain of hemes (He 1, etc.) in four cytochrome molecules, also extremely rapidly (~200ps). [From J. Deisenhofer, H. Michel, and R. Huber, *Trends Biochem. Sci.*, **1985**, 243–248.]

which accepts electrons from the electron-transport chain. Meanwhile, the excited electron reduces an iron-containing protein called *ferrodoxin*, which reduces NADP to NADPH in a process known as *cyclic phosphorylation*. Thus, the coupling together of the two light reactions allows the generation of NADPH and ATP via a two-stage electron pumping mechanism.

3.2.2.4 The generation of oxygen After the excitation of the chlorophyll P680 light-trapping centre of PS2, the excited electron is grabbed by the electron-transport chain and shuttled down to PS1. This leaves the P680 trapping centre minus an electron and in a highly oxidising state. The trapping

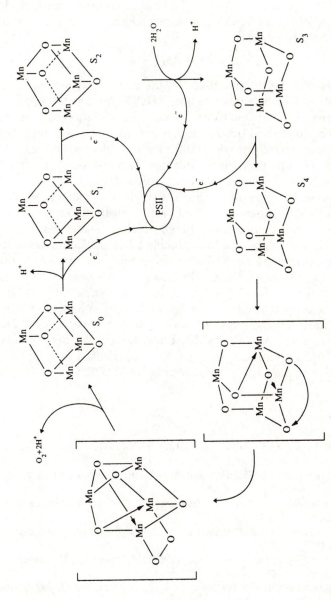

Fig. 3.42 One proposal for the involvement of Mn centres in the photoevolution of dioxygen. [Modified from G.W. Brudvig, and R.H. Crabtree, *Proc. Natl. Acad. Sci. USA*, (1987), **83**, 4856.]

centre's electron deficiency is replenished ultimately by water, but, because of its light-trapping activity, P680 spends most of its time in an electronically excited state. This means that P680 is in constant need of electron replenishment. Water molecules therefore undergo, not a one-, but a four-electron oxidation to oxygen.

$$2H_2O = O_2 + 4H+ + 4e-$$

The trouble is that the intermediates on the way to oxygen, e.g. hydroxyl radicals (HO·) and hydroperoxy radicals (HOO·) are much more powerful oxidising agents than oxygen. Consequently, the oxidation of water to oxygen needs to be conducted quickly and efficiently to ensure that there is no build-up of membrane-destroying (and therefore life-destroying) oxidising intermediates. Nature gets round this problem by using intermediaries between the oxidised P680 and the water molecules.

Splitting of water goes on inside the thylakoid membrane. As P680 becomes oxidised, it replenishes itself from a nearby source of electrons (which may be a plastoquinone complex), which we shall refer to as Z. This is oxidised to Z^+, which in turn immediately takes an electron from another nearby protein complex (called M), possibly containing up to four atoms of the transition metal, manganese. The complex M sequentially loses four electrons this way, at which point it becomes sufficiently oxidised to remove four electrons from two water molecules, releasing oxygen and returning to the uncharged M state ready to do it all over again.[18] The production of dangerous, partially oxidised intermediates is therefore avoided. At the same time protons are generated that are released into the interior of the thylakoids and are used in the generation of NADPH.

3.3 References

1. Coulson, O'Leary, and Mallion (eds) *Hückel Theory for Organic Chemists*, Academic Press, New York (1978).
2. G. M. Badger, *Aromatic Character and Aromaticity*, Cambridge University Press, Cambridge (1969).
3. See, J. March, *Advanced Organic Chemistry*, 4th Edn, John Wiley & Sons, New York (1992).
4. See, J.-H. Führop, in *The Porphyrins*, ed. D. Dolphin, Vol. 2, Academic Press, New York (1978), p. 131.
5. See, R.B. Woodward, *Ind. Chim. Belge.*, (1962), 1293; see also, H. Scheer and H.H. Inhoffen in ref. 4, p. 46.
6. See, M. Gouterman in *The Porphyrins*, ed. D. Dolphin, Vol. 3, Academic Press, New York (1978), p. 1, and references therein.
7. M. Gouterman, *J. Mol. Spectrosc.*, (1961), **6**, 138.
8. P.W. Atkins, *Molecular Quantum mechanics*, Vol. 1, Clarendon Press, Oxford (1970), p. 67.

9. M. Zerner and M. Gouterman, *Theor. Chim. Acta*, (1966), **4**, 44.
10. M.J.S. Dewar and H.C. Longuet-Higgins, *Proc. Phys. Soc.*, (1954), **67a**, 795.
11. See, for example, A. Vincent, *Molecular Symmetry and Group Theory*, John Wiley & Sons, (1977).
12. D.O. Hall and K.K. Rao, *Photosynthesis*, 4th Edn, Edward Arnold, (1987).
13. S.L. Miller and L.E. Orgel, *The Origins of Life on Earth*, Prentice–Hall, New Jersey (1974), p. 33.
14. J. Barber, *Photochem. Photobiol.*, (1979), **29**, 203.
15. D. Mauzerall and F.T. Hong, in *Porphyrins and Metalloporphyrins*, ed. K.M. Smith, Elsevier Scientific Publishing, (1975), p. 703.
16. R.G. Bennett and R.E. Kellogg, in *Progress in Reaction Kinetics*, ed. G. Porter, Vol. 4, Pergammon Press, Oxford (1967), p. 218.
17. (a) See, Th. Förster, in *Modern Quantum Chemistry, Part 3*, ed. O. Sinanoglu, Academic Press, London (1965), p. 93; (b) J. Fajer, *Chem. Ind.*, (1991), 869, and references therein.
18. G.W. Brudvig, H.H. Thorp, and R.H. Crabtree, *Acc. Chem. Res.*, (1991), **24**, 311.

4. How do they do it?—Coping with oxygen

4.1 A brief tutorial on oxygen

4.1.1 What is oxygen?

Oxygen is a gas. Colourless and odourless, without it we suffocate. Yet when its partial reduction products, hydroxyl radicals and peroxy radicals, overwhelm our body's defence systems, oxygen ultimately kills us. It is these partial reduction products that are the most likely cause of ageing.

Oxygen is the most abundant element on earth, making up 47.2% of the earth's crust (in the form of metal oxides, silicates and carbonates, etc.) and 21% of the atmosphere.[1] The other gases are mainly nitrogen, a smattering of carbon dioxide, oxides of nitrogen and sulphur (especially in towns and cities), and the rare gases neon, argon, krypton, and xenon. The most common form of oxygen is diatomic, with the formula O_2. Unlike most diatomic species, the liquid and solid states of oxygen are pale blue in colour and strongly paramagnetic—a clue to oxygen's electronic structure and the key to why oxygen is so important for life. Another allotrope of gaseous oxygen has three oxygen atoms and is called ozone. It is highly reactive, is formed by the action of an electrical discharge, and, more importantly, ultraviolet light on oxygen, a reaction that occurs high in the stratosphere. Ozone itself absorbs UV light, turning back into oxygen. Thus, as oxygen levels built up in the atmosphere of the primeval earth, life-threatening incoming UV radiation was gradually blocked from reaching the surface. Ironically, therefore, as one kind of life was wiped out by the advent of oxygen, so the new life that evolved to take its place was protected from harmful solar UV radiation.

Ozone has continued to protect life on the earth's surface and in its seas from the harmful effects of UV light. Now, however, human activities in releasing chlorofluorocarbons into the atmosphere, are seriously depleting the ozone layer above the earth's two polar regions.[2] As the ozone is destroyed, so more and more UV light penetrates to the earth's surface. Evidence is mounting of the harm caused by increased levels of UV light, e.g. increased incidence of skin cancer and blindness through cataracts in animals as well as humans.

Oxygen itself is a reactive gas and a strong oxidising agent. Anything that burns in air, burns more vigorously in pure oxygen. It is fortunate indeed that the amount of oxygen in the atmosphere is at the level it is. Any lower and complex life forms could not be sustained. Any higher and anything that could burn would have spontaneously gone up in smoke ages ago.[3]

In a way, combustion and cellular metabolism are similar processes. The same amount of energy is released from a gram of sugar regardless of whether it is burnt or metabolised in a cell. The difference is that some of this energy is used by the cell for its own purposes, and it is not all radiated as heat. In both cases, combustion and metabolism, oxygen is the key ingredient. What is it about oxygen that drives these processes? The simple answer is the oxygen molecule's electronic structure.

4.1.2 The electronic structure of diatomic oxygen

Figure 4.1. shows the well-known molecular orbital description of bonding in an oxygen molecule. To generate an oxygen molecule, we bring two oxygen atoms together so that their separate atomic orbitals overlap. This forms bonding and antibonding pairs of molecular orbitals from all the atomic orbitals containing electrons, which are filled with electrons using the Aufbau principle (i.e. the lowest energy orbitals are filled first and each orbital cannot hold more than two electrons). Only the four 2p electrons (i.e. the outermost electrons) from each oxygen atom make significant contributions to the bonding between them. This gives a total of three bonding and three antibonding orbitals in which to distribute eight electrons. The result is the formation of a σ-bond containing two electrons an two π-bonds containing two electrons each. The remaining two electrons each individually occupy a π^*- molecular orbital. The prediction from this simple molecular orbital treatment is that diatomic oxygen in its electronic ground state contains two unpaired electrons and must therefore be paramagnetic. This was one of the first successes for simple MO theory and gives diatomic oxygen a bond order of two $(1\ \sigma + 2\ \pi - [2 \times \frac{1}{2}]\pi^* = 2)$. By adding one electron to each of the half-filled π^* orbitals, the peroxide anion (O_2^{2-}) is formed, and by splitting in two and adding two more electrons, the oxide anion (O^{2-}) with the electron configuration of the noble gas neon $(1s^2 2s^2 2p^6)$, is formed. The two half-filled molecular orbitals on the oxygen molecule are thus the key to its reactivity, providing a "sink" for electrons produced during oxidation–reduction reactions, e.g. during cellular metabolism.

4.2 Oxygen and iron

4.2.1 Rusting

Together, oxygen and iron present a fascinating paradox, the solution to which provides an insight into the molecular basis of several natural

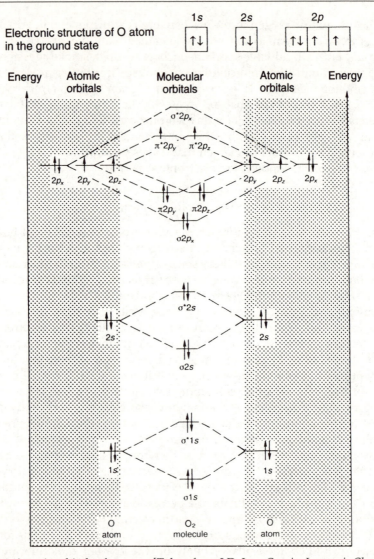

Fig. 4.1 Atomic orbitals of oxygen. [Taken from J.D. Lee, *Concise Inorganic Chemistry*, 4th Edn, Chapman & Hall, London (1991).]

processes. Put simply, the paradox may be presented like this. Iron, in the presence of oxygen and water, turns to rust, the highly insoluble hydrated form of ferric oxide [i.e. Fe(III)]. Anyone who has owned a car for a long period of time, will know that a constant battle has to be fought against rusting bodywork. Yet *our* bodies are full of air, water, and iron-containing proteins, so how come our arteries and cells are not clogged with thick, gelatinous, hydrated ferric oxide? Put another way, if nature had not devised ways of coping with oxygen and iron (and assuming life as we know it were still possible—a ludicrous suggestion) we would each need to drink every day the entire water consumption of New York State in order to shift the rust that would clog our bodies from their metabolic turnover of 15 mg of iron per day.

Rusting is a complex process that is believed to proceed via the ferrous [i.e. Fe(II)] state of iron.[5] When an iron-containing object like a car rusts, a non-coherent layer of hydrated ferric oxide is formed which then flakes away exposing a fresh surface for further attack by air and water.

Ways around rusting are possible. For example, dipping iron in molten zinc or coating by electrolysis affords some protection. Alternatively, rusted bodywork can be painted with red lead (a process known as Parkerising) or phosphoric acid (this creates a protective layer of iron phosphate and is called Bonderizing). But the point to remember is that the rusting process, i.e. the oxidation of iron to hydrated oxide, is irreversible.

Rusting may give a clue as to why oxygen had such a disastrous effect on the anaerobes when it first appeared. Oxygen levels in the atmosphere probably did not build up gradually over eons. It is far more likely that the atmosphere transformed from about 1% oxygen to 17–21% overnight, geologically speaking.

The first photosynthetic bacteria, some 3.5–4 billion years ago, were restricted to hydrogen sulphide as a source of hydrogen atoms for energy transfer. They obtained their H_2S from the sea around underwater volcano vents. The atmosphere was then reducing, so that UV light was free to bombard land surfaces. With no oxygen in the atmosphere, it also meant that any iron would have been in the more water-soluble Fe(II) state. The first blue-green algae, which appeared around 2.5 billion years ago, had learnt the trick of using more abundant water as a source of reducing equivalents for photosynthetic energy transfer. They started to produce oxygen, but instead of it going into the atmosphere, the first oxygen would have combined with Fe(II) to produce red precipitates of ferric oxide. So, even though blue-green algae were producing oxygen, atmospheric concentrations of the gas would have remained around 1% until all the soluble ferrous salts had been precipitated out. Around 1.7 billion years ago, with the last of the water-soluble Fe(II) precipitated out of the seas, atmospheric O_2 levels suddenly shot up. The era of domination by the anaerobes was over.

Since then, living things have evolved that have learnt to cope with and exploit an oxygenated atmosphere. For example, many proteins contain iron; yet, the iron moieties are not locked into an Fe(III) oxidation state. Far from it, nature has chosen iron because it is ideal for the electron-transfer reactions that are the metabolic basis of life. Iron is abundant, and it exhibits more than one oxidation state. In the presence of air and water, iron prefers to be in the Fe(III) state, forming hydrated ferric oxide.

How then does nature ensure that iron performs its biochemistry and does not simply precipitate out of living things? The answer is packaging. Nature takes special care to wrap iron in luxurious water-soluble protein coats. There are several types of iron-containing proteins. Two important groups—the iron storage and transporting proteins respectively—store the iron until it is required (ferritins) and transport it to where it is needed (siderophores), thus ensuring that it does not clog our blood vessels and cells. The group we shall be concentrating on though, are called hemoproteins. As their name implies, these are proteins in which the iron atom [in the Fe(II) oxidation state] is complexed to a protoporphyrin IX moiety to give heme. We shall be examining a few of these. In particular, we shall concentrate on probably the most important hemoprotein of all as far as large vertebrates are concerned: the red, oxygen-carrying pigment in blood, hemoglobin. Firstly, though, we shall briefly review some salient points of iron chemistry.

4.2.2 The chemistry of iron

Iron, along with the heavier elements from the periodic table, is made from the lighter elements inside stars by thermonuclear fusion just before the star explodes in a supernova explosion. This scatters iron and the other elements deep into space, where they mix with interstellar gas clouds that eventually form new stars and planets.

Iron is the fourth most abundant element in the earth's crust (5%) and the second most abundant metal, being present as the ores hematite (ferric oxide, Fe_2O_3), limonite (hydrated ferric oxide), magnetite (Fe_3O_4), siderite (ferrous carbonate), iron pyrites (FeS_2), chromite ($FeCr_2O_4$), and chalcopyrite ($CuFeS_2$).[5] In order to extract it, the iron is roasted in air to remove the impurities, and then reduced to the metal with carbon in a blast furnace to give cast iron.

Iron has an atomic weight of 55.85, the fractional number being due to the presence of different iron isotopes (^{54}Fe = 5.84%; ^{56}Fe = 91.68%; ^{57}Fe = 2.17%; ^{58}Fe = 0.31%) and belongs to the first row of d block elements, or the transition series.

4.2.2.1 Iron as a transition element Why is iron a transition metal? To understand this we need to look at the periodic table of elements, where iron occupies a central position in the first row of d block elements. The properties of these elements are *transitional* between the metallic behaviour of the s-block elements and the variable valency of the p-block. But the variable valency of the transition metals is entirely different to that of the p-block elements. Whereas the latter have valencies that increase in steps of

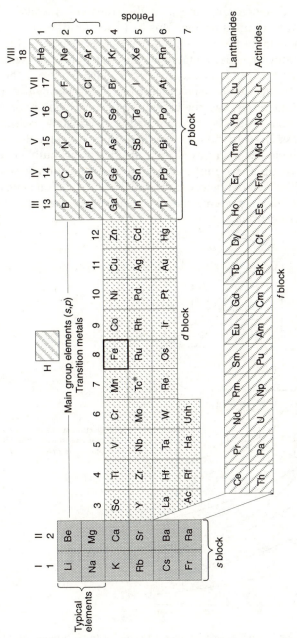

Fig. 4.2 Shape of the periodic table and the place of iron.

two [e.g. tin exhibits Sn(II) and Sn(IV) oxidation states, and arsenic exhibits As(III) and As(V) oxidation states], in the transition metals, changes in valency are always in units of one. This behaviour is due to the way the d orbitals change their energy as we move across the periodic table.

Before scandium (the first element of the first transition series), the 3d energy level is virtually constant in energy, shielded from the growing nuclear charge by the inner shells of electrons. However, with the onset of the first transitions series, and the beginning of the filling of the 3d orbital, it is as if the latter suddenly wake up to the electrostatic presence of the nucleus and take a steep dive in energy.[6] Very quickly, the 3d and 4s orbitals achieve similar energies so that an increasing number of valence possibilities open up as we move across the first transition series, peaking at manganese (the element before iron) which has oxidation states from -1 to $+7$. This last state represents maximum use of all manganese's valence 3d (5) and 4s (2) electrons.

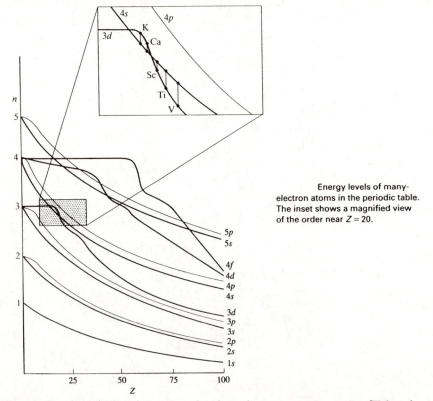

Energy levels of many-electron atoms in the periodic table. The inset shows a magnified view of the order near $Z = 20$.

Fig. 4.3 Overlap of 3d and 4s energy levels in the 1st transition series. [Taken from D.F. Shriver, P.W. Atkins, and C.H. Langford, *Inorganic Chemistry*, Oxford University Press, Oxford (1990), p. 23.]

Iron cannot use all its 3d electrons. The iron atom has an electron configuration of $[Ar]3d^6 4s^2$. As the outer electrons are removed, the remaining electrons are exposed to more of the nuclear charge, which means that their energy declines and they are pulled closer to the nucleus. A moment is reached when no more can be removed as the 3d level has contracted too far for them to be easily got at. Manganese represents the last element in the first transition series in which the 3d level is of high enough energy for all of the 3d electrons to be used in bonding. And so iron exhibits fewer oxidation states than manganese (6 instead of 9; -2, $+1$, $+2$, $+3$, $+4$, and $+6$) of which $+2$ and $+3$ are the most common. This corresponds to electron configurations $[Ar]3d^6$ and $[Ar]3d^5$, respectively.

4.2.2.2 Iron as part of a complex   In general, Fe(II) compounds are relatively easy to oxidise to Fe(III) compounds: the rusting of the metal produces ferric not ferrous oxide. Yet this instability of Fe(II) compounds with respect to Fe(III) is strongly dependent on the ligands surrounding the metal cation. In some circumstances, Fe(II) compounds can be quite stable with respect to oxidation to Fe(III). This is because the ligands affect the relative energies of the 3d orbitals with respect to one another by the well-known process of ligand field splitting.[7] The simplest description of this process assumes that the only interaction between the electrons of the ligand and those of the iron, is electrostatic.

In a hypothetical, isolated Fe(II) cation, the five 3d orbitals all have the same energy (i.e. they are degenerate). Surrounding the Fe(II) cation with six ligands in an octahedral configuration causes a split in the degeneracy of the five 3d orbitals. Two of the 3d orbitals (the ones pointing directly at the approaching ligands, along the x, y, and z axes, i.e. the d_{z^2} and the $d_{x^2-y^2}$) are lifted together in energy above the other three 3d orbitals (which point between the directions of the approaching ligands, and are the d_{xy}, d_{xz}, and the d_{yz} orbitals). The six 3d electrons of the Fe(II) cation can now be distributed between the 3d orbitals. Only if the ligand field splitting of the degeneracy of the 3d orbitals is small will this distribution occur according to Hund's rule (i.e. electrons will tend to have parallel spins and have lower energy, so they will try not to pair up, preferring to occupy a set of degenerate orbitals singly). Under these circumstances, the Fe(II) cation has four of its 3d electrons unpaired, (i.e. in a high-spin configuration) and is paramagnetic.

If, on the other hand, the splitting of the degeneracy of the 3d orbitals is large, then the energy difference between the two sets of 3d orbitals will now be such as to overcome the electrons' propensity for solitude, and they will pair up. Now, the Fe(II) cation is in a low-spin state with no unpaired electrons, and is diamagnetic. This symmetrical arrangement of electrons in the low-spin $3d^6$ state is particularly stable, so that under these circumstances, Fe(II) cations are more difficult to oxidise to Fe(III) than in the high-spin $3d^6$ state.

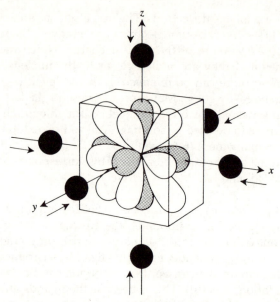

Fig. 4.4 Six ligands approaching a transition metal cation to provide an octahedral crystal field. [Taken from James E. Huheey, E.A. Keiler, and R.L. Keiler, Harper-Collins, *Inorganic Chemistry; Principles of Structures and Reactivity* 4th Edn.]

This is the simplest way of describing how the iron atom's electrons are affected by the surrounding ligands. it is not necessarily the best way (molecular orbital theory does a much better job) as it sometimes yields false conclusions about which ligands should or should not have a powerful splitting effect on the five 3d orbitals. Nevertheless, it provides a qualitatively simple picture which is good enough for our purposes.

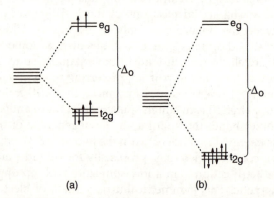

Fig. 4.5 Electron configurations of (a) a d^6 ion in a weak octahedral field and (b) a d^6 ion in a strong octahedral field.

4.2.3 Iron in hemoglobin and myoglobin

In hemoglobin and myoglobin, the Fe(II) cation is coordinated to the four nitrogens of a protoporphyrin IX ligand. This accounts for the ligands in the *x*- and *y*-plane in our simplified example above. The fifth and sixth positions (corresponding to the two ligands approaching the Fe(II) cation along the *z*-axis) are occupied by two imidazole ligands (which are the side chain residues of histidine amino acids in the protein chain wrapped around the heme moiety), one closer to the Fe(II) cation than the other. This allows space for an oxygen molecule to squeeze between one of the imidazole groups and the Fe(II) cation and coordinate to the latter. Interestingly, without the molecule of oxygen, the Fe(II) cations of deoxyhemoglobin and deoxymyoglobin are in a high-spin state and the molecules are paramagnetic overall. When an oxygen molecule coordinates to the Fe(II) cation, the latter's 3d electrons air up to produce a low-spin Fe(II) cation which is diamagnetic and difficult to oxidise to Fe(III).

Another point worth bearing in mind is that most of the ligands binding to the Fe(II) cation in these hemoproteins all contain nitrogen. Now, nitrogen-containing ligands have a greater effect on the splitting of the degeneracy of the Fe(II) 3d orbitals than ligands containing oxygen atoms, such as water (nitrogen is less electronegative than oxygen, so that its orbitals extend further into space, and interact more strongly with metal d orbitals). Water, however, has a great affinity for iron. If water were to get into the heme-containing cavity in the peptide matrix of hemoglobin and myoglobin, it would compete successfully with the imidazole groups for binding to the Fe(II) cation. This would ensure that the Fe(II) cation would remain in the high-spin state and so would be easily oxidised to Fe(III) when it coordinated to oxygen. In hemoglobin and myoglobin, this problem is overcome by ensuring that the heme-containing cavity is lined with hydrophobic amino acid side chains which effectively exclude water.

4.3 Myoglobin and hemoglobin

4.3.1 Oxygen uptake and release

Myoglobin stores oxygen in muscle tissue, while hemoglobin transports oxygen from the lungs to the cells, then picks up carbon dioxide and transports it back to the lungs for exhalation. We can begin to understand how this happens at a molecular level by examining the oxygen uptake characteristics of the two hemoproteins.[8]

This is done by plotting the percentage of oxygenated hemoprotein as a function of oxygen partial pressure. For myoglobin, the curve obtained fits that described by a rectangular hyperbola (see Figure 4.6). This is precisely

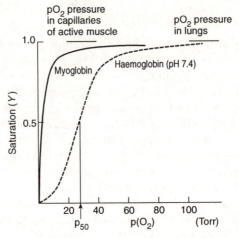

Fig. 4.6 Oxygen-binding curves for myoglobin and hemoglobin.

the kind of relationship that would be expected for a simple bimolecular equilibrium represented by the stoichiometric equation between oxygen and myoglobin as:-

$$Mb + O_2 = MbO_2 \; (Mb = myoglobin)$$

When a similar plot of percentage oxygenated hemoprotein against oxygen partial pressure is performed on hemeoglobin, a completely different picture emerges. What we see is an elongated S-shaped graph, called a sigmoid curve. Unlike the rectangular hyperbolic curve for myoglobin, the sigmoid relationship demonstrates that the affinity of hemoglobin for oxygen increases as oxygen is taken up. Another important point is that these curves are reversible; for hemoglobin, this means that its affinity for oxygen falls as the oxygen is given up. These characteristics have profound physiological significance for they ensure that there is maximum efficiency of uptake, transport, and release of oxygen to the tissues. They also point to differences in the way myoglobin and hemoglobin function, and this, of course is related to their structures; for hemoglobin is a much more sophisticated molecule than myoglobin.

4.3.2 Structure and function of myoglobin and hemoglobin

Hemoglobin is approximately four times the size of myoglobin, the former consisting, more or less, of four myoglobin units. So, the first large difference between the two hemoproteins is that myoglobin contains one heme moiety compared with hemoglobin's four.

In myoglobin, the protein part of the molecule consists of a single chain of 152 amino acids in a genetically determined sequence. In hemoglobin, there are four globin chains, two of them containing 141 amino acids and called α chains, and two containing 146 amino acids and called β chains. The protein chains of myoglobin and hemoglobin are formed into varying lengths of helical and non-helical segments. In these hemoproteins, 70–80% of the amino acids form helices but most proteins have much less helical structure; in cytochrome *c*, for example, only 30% of the amino acids form helices.

The helix is like a spiral staircase arranged in such a way that identical parts of amino acids do not lie above each other. Thus, each amino acid contributes a step 1.5 Å in height so that 3.6 steps are required to make a complete turn of the screw. Such a turn gives an increase in height of 5.4 Å. Hydrogen bonding between the imino (—NH—) hydrogen of one amino acid and the carbonyl ($>C=O$) group of another amino acid above or below it, stabilise the helix.[9]

Each helical region of a globin chain is labelled with a capital letter, running alphabetically from the N-terminus to the C-terminus. Thus, the letter A would denote the first helix from the N-terminus, while H denotes the C-terminal helix. Non-helical regions at the N-terminus are denoted as NA, those at the C-terminus as HC. Non-helical regions between the helices are denoted as AB, BC, etc. So it is now possible to pinpoint exactly any amino acid in any protein chain of hemoglobin and myoglobin by reference to:-

(a) its chemical abbreviation (e.g., "gly" is glycine, "his" is histidine, etc.);

(b) in the case of hemoglobin, which globin subunit it is part of, e.g. α_1, α_2, β_1, or β_2;

(c) which helical or non-helical region it belongs to;

(d) where in the helical or non-helical region it belongs;

(e) finally, its position in the whole amino acid sequence, numbering from the N-terminus.

Thus His β, F8 (92) refers to a histidine amino acid residue in a β-globin chain, situated in the eighth position of the F helix (from the N-terminus) and in position 92 of the whole 146 amino acid sequence.

To the casual observer, presented with a picture of hemoglobin's serpentine polypeptide coils, it seems as if there is no rhyme or reason to the mass of helices and non-helical regions. However, the sequence of amino acids has been genetically determined and study of the coiling of the globin chains shows that the whole structure forms a basket-like environment around each heme, which is therefore in loose contact with about sixty different atoms in the protein. What is more, these sixty atoms belong to neutral, hydrophobic amino acid side chains—the polar amino acid side chains occupy positions on the outside of the globin molecules. Protein molecules

(a)

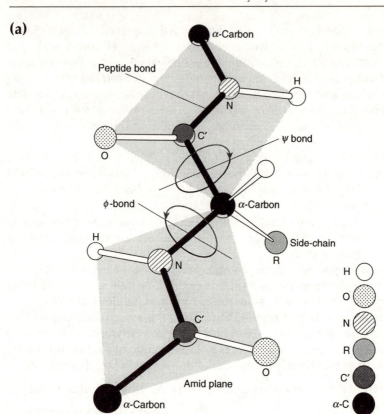

(c)

(b)

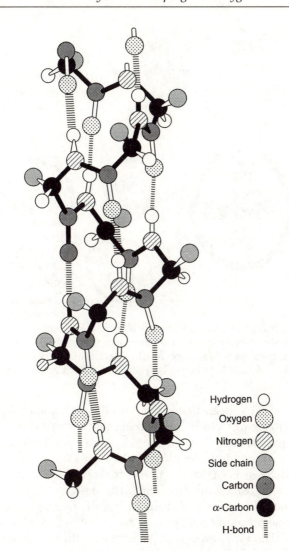

Hydrogen ◯
Oxygen ⬚
Nitrogen ⬚
Side chain ◯
Carbon ⬤
α-Carbon ⬤
H-bond ⦂

Fig. 4.7 (a) Polypeptide chain showing φ, ψ, and peptide bond for residue R₁ with chain. (b) An α helix. (c) β-Pleated sheet structure between two polypeptide chains.

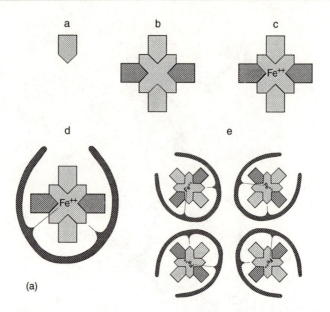

Fig. 4.8 (a) A cartoon of steps in the development of hemoglobin: a pyrrole; b porphyrin; c heme; d heme–globin complex (e.g. myoglobin); e four heme-globin units in one complex (e.g. hemoglobin). (b) Quaternary structure of hemoglobin. The $\alpha_1\beta_2$ interface contacts between FG corners and C helix are shown.

like the globins, have been likened to an oil drum filled with oil. This is because the outer part (which corresponds to the metal drum) is wettable, but the inner part is, like oil, water repelling and bound to other hydrophobic parts of the molecule by weak van der Waals forces. In addition, each heme is linked, by its Fe(II) cation, to the imidazole side chain of a histidine amino acid residue, called the proximal histidine.

Although they are embedded deep within the hydrophobic heart of the globin, the heme propionic acid groups reach out to the polar, external side of the protein. The vinyl groups make contact with the globin's non-polar interior. The four globin subunits of hemoglobin give the overall molecule a roughly spherical shape, and are arranged tetrahedrally around a water-filled cavity, which coincides with a twofold axis of symmetry.

We can now return to hemoglobin's sigmoidal oxygen uptake curve. The fact that it differs so markedly from myoglobin's rectangular hyperbolic uptake curve suggests that hemoglobin is not behaving simply as a "supermyoglobin", with each hemoglobin subunit acting independently of the other three. Far from it: the fact that the hemoglobin's affinity for oxygen increases as oxygen is taken up suggests that the hemoglobin subunits are cooperating with each other in some way. This is called the *cooperative*

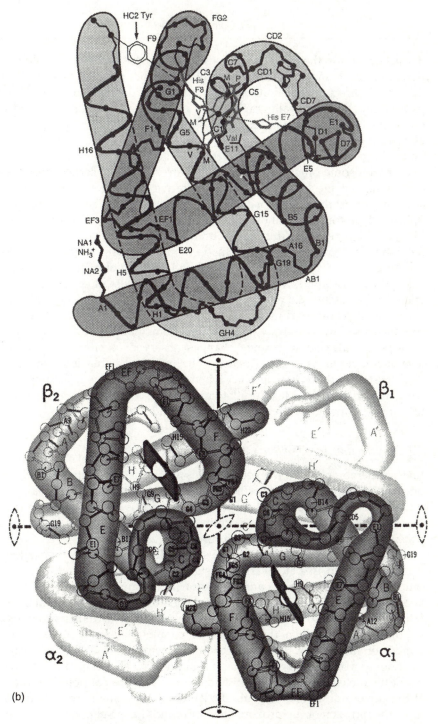

Fig. 4.9 Heme embedded in basket-like globin.

interaction, and in the early days of research into the way hemoglobin functions, the mechanism of this cooperation was thought, erroneously, to be
due to a direct interaction between the individual heme units.[10] The
painstaking X-ray crystallographic data of Max Perutz finally finished off
the heme–heme interaction theory by showing that the closest approach of
any two heme units (in the α, and α_2 chains) is never less than 25 Å. This is
far too large for there to be any direct interaction.

There are some other observations, however, that any serious attempt at
explaining the mechanism of the cooperative interaction would have to
address. For example, there is the so-called Bohr effect. This is the observed
uptake of protons by hemoglobin as it releases oxygen. In other words,
there is a change in pH in the surrounding medium as we go from oxyhemoglobin to deoxyhemoglobin. This is thought to help in the removal of
carbon dioxide, which binds to the tail-end of the globin chains. It is also
thought that the uptake of protons catalyses the release of oxygen to tissues
made acid by lactic acid and bicarbonate. An organic phosphate, called 2,3-
diphosphoglycerate and manufactured in red cells as a response to oxygen
shortage, also lowers the oxygen affinity of hemoglobin.

There are also some strange changes observed *in vitro*. For example,
deoxyhemoglobin is paramagnetic, while oxyhemoglobin is diamagnetic.
And crystals of deoxyhemoglobin, when exposed to oxygen, shatter. It was
left to the brilliance of Max Perutz to give the first truly coherent interpretation of the cooperative interaction in atomic and molecular terms.[11]

4.3.3 Molecular mechanism of the cooperative interaction

In deoxyhemoglobin, all the Fe(II) cations are in the high-spin d^6 configuration, which is paramagnetic with four unpaired electrons. This means that
they are just too big to fit into the hole in the middle of the porphyrin rings.
The Fe(II) cations sit above the porphyrin ring planes, directly linked to the
proximal histidine as shown in Figure 4.10.

When oxygen binds to a heme high-spin Fe(II) cation in deoxyhemoglobin,
it is loosely held in place via the distal histidine, and converts the Fe(II)
cation into the low-spin d^6 configuration with no unpaired electrons, and so
is diamagnetic. This explains why deoxyhemoglobin is paramagnetic and
fully oxygenated oxhyemoglobin is diamagnetic. But there is a much more
important effect, which this change in paramagnetism is merely symp-
tomatic of; that is, that the pairing of the d electrons causes the Fe(II) cation
to undergo a shrinkage in volume of around 13%. This is because the pairing of the d electrons places them in d orbitals that are not pointing directly
at the surrounding ligands of the Fe(II) cation (the four porphyrinic nitrogens, the nitrogen of the proximal histidine and the oxygen molecule). They
become less shielded from the nuclear charge of the cation and are pulled

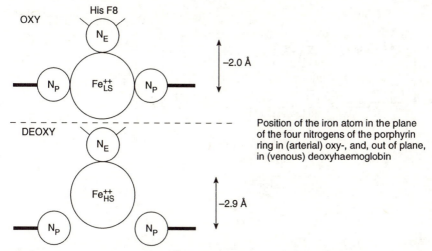

OXY

His F8

N_E

N_P Fe_{LS}^{++} N_P

−2.0 Å

DEOXY

N_E

Fe_{HS}^{++}

N_P N_P

−2.9 Å

Position of the iron atom in the plane
of the four nitrogens of the porphyrin
ring in (arterial) oxy-, and, out of plane,
in (venous) deoxyhaemoglobin

Fig. 4.10 High- and low-spin d^6 Fe(II) cation in porphyrin bound to proximal histidine (N_ϵ).

in, effectively decreasing the size of the Fe(II) cation. Because of its shrinkage in size, the Fe(II) cation will now fit snuggly into the hole in the middle of the porphyrin, resulting in a motion of the Fe(II) cation of about 0.5–1.0 Å. The Fe(II) cation is connected to an imidazole group, so that the movement of the iron leads to a larger motion of the attached proximal histidine residue. This motion triggers a series of structural changes, which start in a particular globin subunit, but which goes on to affect all of them. Perutz coined the term "molecular amplifier" to describe the action of the porphyrin in converting the small shrinkage of the Fe(II) cation into the large motion of the proximal histidine.

The four globin units in deoxyhemoglobin are held together by hydrogen bonding between amino acid residues, sometimes on neighbouring α and β-chains, sometimes on the same chain. This is on top of the hydrogen bonding previously described that stabilises the helical structures in the globin chains. Thus, the C-terminal amino acid in the α chains is arginine, which in deoxyhemoglobin forms two hydrogen bonds with a polar side chain on amino acids in the opposite α chain. Similarly, the C-terminal amino acid of the β chains is histidine, and this forms two hydrogen bonds, one with a polar side chain of an amino acid in the same β chain, and another with a polar side chain of an amino acid in a neighbouring α chain. Other amino acid side chains are involved in this hydrogen bonding and a more complete (but diagrammatic) picture is shown in Figure 4.11(b).

Now, a hydrogen bond has, on average, only about one-tenth of the binding energy of a covalent bond. In small molecules, therefore, (where the total binding energy is dominated by covalent bonding) hydrogen bonding

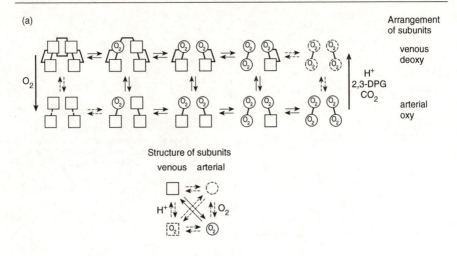

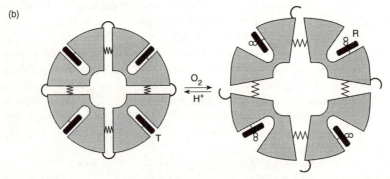

Fig. 4.11 (a) Cartoon of structural changes between deoxy- and oxy-hemoglobin. (b) Another way of viewing the hemoglobin–O_2 interaction.

tends to make little difference to the overall structure of the molecule. Even so, in water, hydrogen bonding has a profound effect on the physical and chemical properties of this versatile liquid, so necessary for life. For example, compare water, an odourless and tasteless liquid at room temperature, with (under the same circumstances) the foul-smelling gas hydrogen sulphide (oxygen and sulphur are in the same group in the periodic table). So, even in small molecules, hydrogen bonding plays an important part in determining the overall properties of the compound.

In a macromolecule like hemoglobin, because of the larger number of amino acids with polar side chains in its structure, there are many opportunities for hydrogen bonding. This means that hydrogen bonding can account for a sizeable chunk of the molecule's binding energy, and so make an important contribution to determining its structure. Perutz's work in determining the structures of deoxy- and oxyhemoglobin makes it crystal clear how the breaking of hydrogen bonds as hemoglobin becomes oxygenated, turn hemoglobin from a tight, almost "spring-loaded" hydrogen-bonded structure (with low affinity for oxygen), into a much looser, more open structure, as it becomes oxygenated. It is almost as if, in molecular parody of the lungs it passes through on its journey around the body, hemoglobin "breathes in" as it picks up oxygen, expanding in size, and "breathes out" or contracts as it gives up its precious load of oxygen to the cells that it serves.

The penultimate amino acid in each of hemoglobin's four globin chains is tyrosine; Tyr α, HC2(140) for the α-chains, and Tyr β, HC2(145) for the β chains. Tyrosine has a phenol side chain, which, in deoxyhemoglobin, fits into a cleft between the F and H helices. On uptake of oxygen, the Fe(II) cation shrinks (as described earlier) pulling down (rather like an old-style lavatory chain) on the proximal histidine. This motion moves the F helix closer to the H helix, so that the tyrosine-containing pocket becomes too small and squeezes out the phenolic side chain.

Now, the C-terminal amino acids (the arginines in the α chains and the histidines in the β chains), which form hydrogen bonds that hold the chains in the "spring-loaded" deoxyhemoglobin structure, are attached directly to these tyrosines. Their motion causes the hydrogen bonds to break, not all at once but in a stepwise manner. It is the Fe(II) cations of the α chains that are oxygenated first, so it is their hydrogen bonds that break first. While intact,

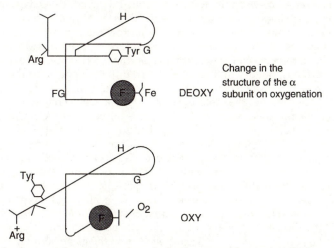

Fig. 4.12 As heme takes up oxygen, it leads to conformational changes of the attached globin protein.

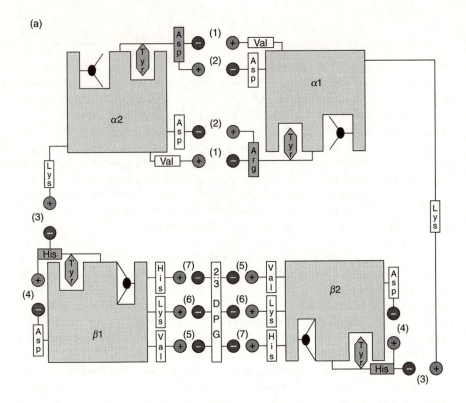

Fig. 4.13 (a) Perutz's concept of the deoxygenated tetramer. The α heme pockets are ready to receive oxygen molecules, the penultimate tyrosines of both α and β chains are tethered in pockets between the F and H helices and a 2:3 diphosphoglycerate molecule is inserted between the widely separated β chains. (b) Oxygenation of the α hemes. First one and then the other α heme has been oxygenated. The iron atoms moving into the plane of the porphyrin ring narrow the gap between the F and H helices and as a result the penultimate tyrosines are expelled from their niches. At this stage the links between the α C-terminal arginine, 141 (HC3), with the N-terminal valine 1 (NA1), bond 1 and aspartic acid 126 (H9), bond 2, of the opposite α chain are broken. (c) Movement into the oxygenated conformation. The β chains have moved together and the molecule of 2:3 diphosphoglycerate which previously lay between them has been expelled. The bonds (3) between the β C-terminal histidines 146 (HC3) and Lys 40 (C5) of the opposite α chain are broken, weakening the tether holding the β C-terminal histidines in position. It is not necessary to assume that the major conformation change between stage B and C will only occur when both α chains are oxygenated and both β chains are deoxygenated. It is possible but probably less likely, that it could also occur earlier (i.e. when one α heme is oxygenated, or later when 2α and one β heme are oxygenated). (d) Oxygenation of the β hemes. The bonds between the β C-terminal histidines 146 (HC3) and the aspartic acid 94 (FG1) in their own polypeptide chain (bond 4) rupture. The heme pocket widens as the penultimate tyrosine is expelled from the niche between the F and H helices. Oxygen is now able to enter the β heme pockets and oxygenation of the hemoglobin tetramer is complete.

(b)

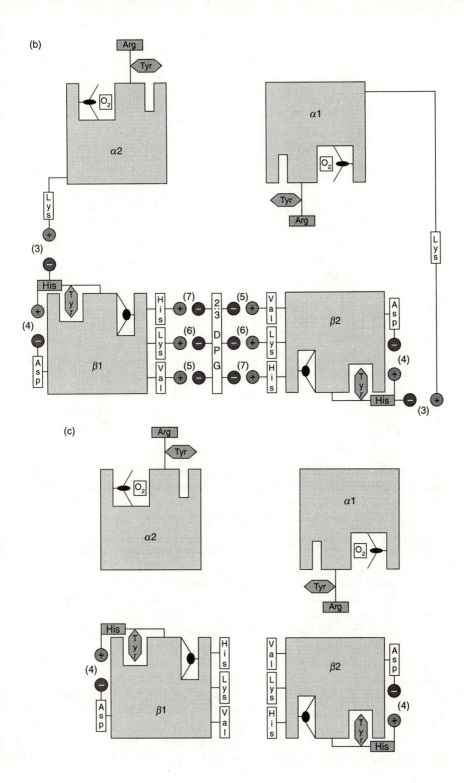

(c)

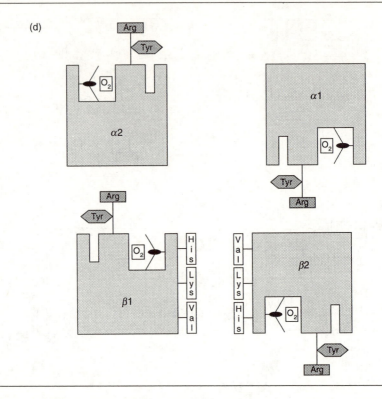

however, the hydrogen bonds keep hemoglobin in its low oxygen affinity, high-spin form. As they are broken (and protons liberated), so the molecular equilibrium between the "spring-loaded" deoxy and the "relaxed" oxy forms of hemoglobin moves in favour of the latter. At any stage in the process of oxygen uptake, the equilibrium may click over to the relaxed form, with the chances of this happening increasing as oxygen is taken up (see Figure 4.11.).

The change from the more compact, spring-loaded deoxy form to the looser, relaxed oxy form explains why crystals of deoxyhemoglobin shatter as they take up oxygen. The breaking of hydrogen bonds on oxygenation (and their reformation on deoxygenation) explains the release and uptake of protons behind the Bohr effect.

The role of 2,3-diphosphoglycerate (DPG) is also explained by this mechanism. It forms four extra hydrogen bonds between itself and the β chains. On conversion to the oxy form of hemoglobin, the N-termini move apart and the H helices of the β chains move closer together, forcing the DPG out. Conversely, when oxygen is being given up (a process that is assisted by protons, especially in cells with low oxygen content) DPG locks the β chains back into the deoxy configuration, so assisting their release of oxygen. In

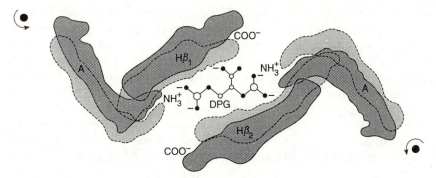

Fig. 4.14 The DPG molecule has been tentatively placed in the central cavity, centred approximately on the twofold axis. The picture shows that is phosphates are close to the N-termini of the β chains in deoxyhemoglobin. On oxygenation the N-termini move apart and the H-helices close up, thus expelling the DPG.

the presence of protons, therefore, DPG catalyses the release of oxygen from oxhyemoglobin.

Carbon dioxide competes with protons and DPG for binding to hemoglobin. In fact, hemoglobin carries 60% of the carbon dioxide produced during metabolism back to the lungs. It does this by reacting with the N-terminal -NH_2 groups to give the carbamate,

$$R—NH_3 + = R—NH_2 + H+$$
$$R—NH_2 + CO_2 = R—NH—COOH$$

which is highly labile to decarboxylation, a process that goes on in the lungs.

Later, we shall see how genetic disorders lead to changes in the amino acid sequence of the globin around the heme site. Such structural changes can allow oxygen to oxidise the Fe(II) cation to Fe(III). Under these circumstances, methemoglobin (as hemoglobin with oxidised heme centres is called) is incapable of reversibly binding oxygen. However, at this point it is only necessary to observe how a change in protein structure [which changes the electron-donating or -withdrawing nature of the amino acids that bind to the heme Fe(II) cation] can completely alter the redox chemistry of the heme unit. Apart from crippling the oxygen-transporting role of hemoglobin, it also leads to a wide variety of hemoproteins with different functions.

4.4 Hemoproteins in cell metabolism

4.4.1 Heme without the protein—the chemistry of heme on its own

The alliance between heme and a protein, to make a hemoprotein, completely alters the chemistry of the heme. If we are to appreciate how, then we need to know how heme, Fe(II) protoporphyrin IX, behaves on its own.

Redox processes, i.e. electron transfer, can occur at the metal centre, via axial processes, or at one of the porphyrin *meso*-positions, via peripheral processes.[12] Unlike a penta- or hexa-coordinated iron complex (with a coordination sphere of the same five or six monodentate ligands), the hemes (and hemins) have their iron cations more strongly bound to the equatorial tetradentate porphyrin than to the axial ligands. Two kinds of axial redox process are possible. In the first case, an axial process can involve direct charge transfer (akin to the outer sphere mechanism of inorganic chemistry[13]) through an axial ligand (see Figure 4.15). All that is required for

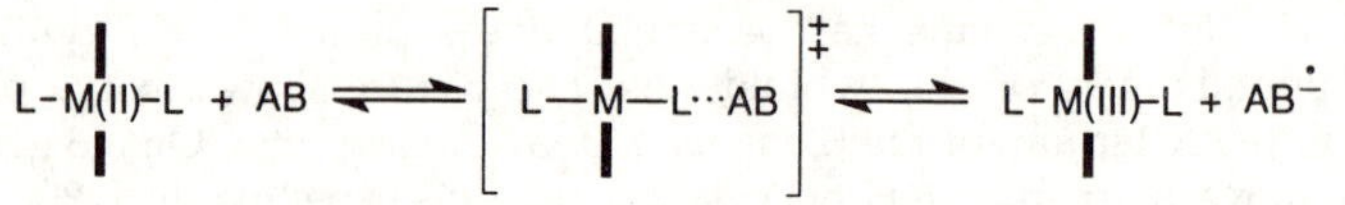

$$L-M(II)-L \; + \; AB \; \rightleftharpoons \; \left[L-M-L\cdots AB \right]^{\ddagger} \; \rightleftharpoons \; L-M(III)-L \; + \; AB^{\overset{\cdot}{-}}$$

Fig. 4.15 Axial through-ligand (outer sphere) mechanism for metalloporphyrin redox reactions.

electron transfer to take place is for the oxidant or reductant to be in orbital overlap with the axial ligand. Alternatively, a redox process may involve the oxidant or reductant becoming an axial ligand of the metal. This can necessitate ligand exchange of one of the axial ligands, or simple addition of the reactant to an axial position (akin to the inner sphere mechanism of inorganic chemistry). In the second case, peripheral processes involve oxidation or reduction at the *meso*-position. The aromatic electronic configuration of the porphyrin α-system stabilises a α-cation (oxidation) or

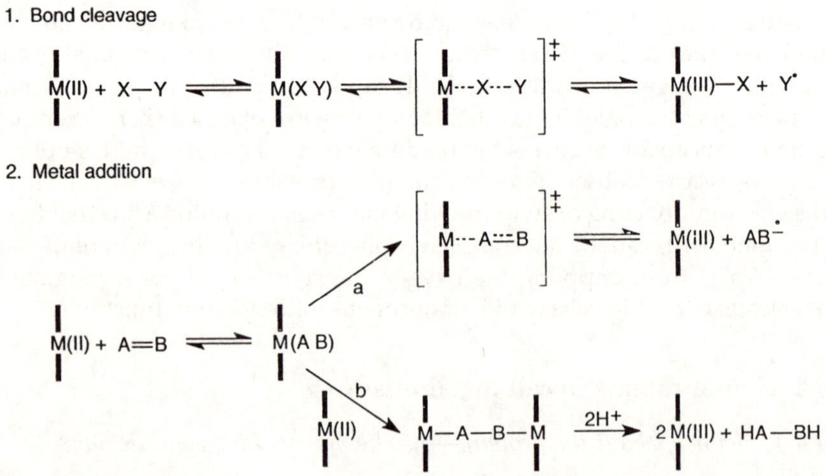

1. Bond cleavage

$$M(II) \; + \; X-Y \; \rightleftharpoons \; M(X\,Y) \; \rightleftharpoons \; \left[M\cdots X\cdots Y \right]^{\ddagger} \; \rightleftharpoons \; M(III)-X \; + \; Y^{\cdot}$$

2. Metal addition

$$\left[M\cdots A\overset{\cdots}{=}B \right]^{\ddagger} \; \rightleftharpoons \; M(III) \; + \; AB^{\overset{\cdot}{-}}$$

$$M(II) \; + \; A{=}B \; \rightleftharpoons \; M(A\,B)$$

$$M(II) \;\diagdown\; M-A-B-M \; \xrightarrow{2H^{+}} \; 2\,M(III) \; + \; HA-BH$$

Fig. 4.16 Axial bond cleavage (inner sphere) mechanisms for metalloporphyrin redox reactions.

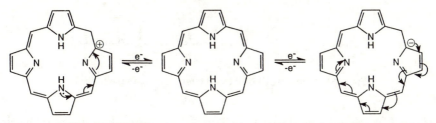

Fig. 4.17 Stabilisation of porphyrin π-cations and -anions.

α-anion (reduction) radical long enough for internal charge transfer with the metal centre to occur.

Two distinct peripheral processes can be identified, once again mirroring the outer and inner sphere mechanisms just mentioned. The first involves simple orbital overlap between the reactant's π-system and that of the porphyrin near the *meso*-position. No σ-bonds are formed or broken, only electron transfer (i.e. an outer sphere process) takes place, and as such is likely to be reversible. The second process involves transfer of an atom along with an electron (i.e. an inner sphere process). The species can be hydrogen or a free radical. Transfer is likely to be irreversible and involves formation and breaking of σ-bonds at the *meso*-position. Both of these peripheral processes require the metal to be in conjugation with the porphyrin π-system for oxidation of the metal to take place. This happens

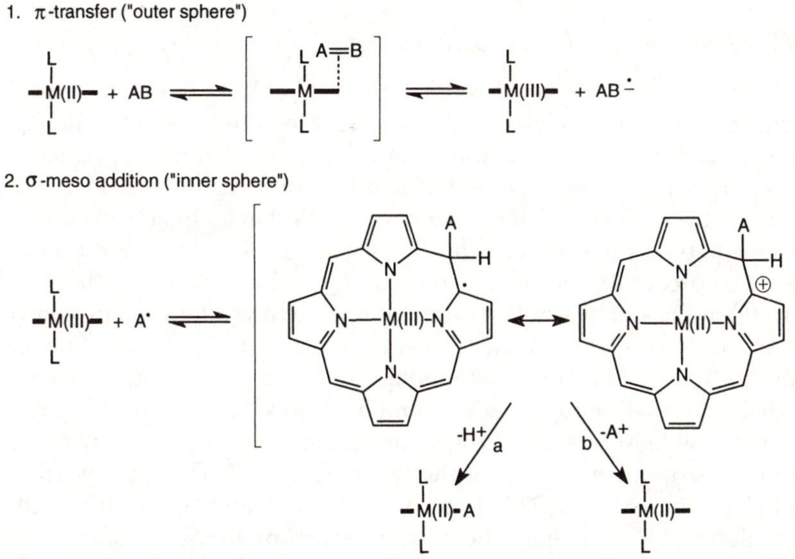

1. π-transfer ("outer sphere")

$$-M(II)- + AB \rightleftharpoons \left[-M- \overset{A=B}{\underset{\vdots}{}} \right] \rightleftharpoons -M(III)- + AB^{\cdot-}$$

2. σ-meso addition ("inner sphere")

Fig. 4.18 Peripheral 'outer sphere' and 'inner sphere' mechanisms for metalloporphyrin redox.

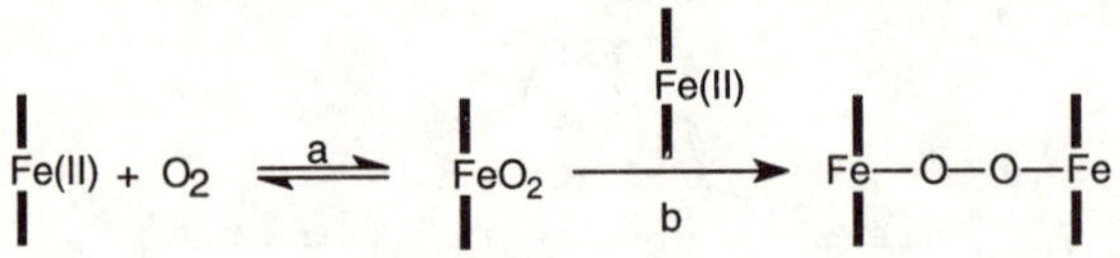

$$\text{Fe(II)} + \text{O}_2 \underset{a}{\rightleftharpoons} \text{FeO}_2 \xrightarrow{b} \overset{\displaystyle \text{Fe(II)}}{\underset{\displaystyle \text{Fe}-\text{O}-\text{O}-\text{Fe}}{}}$$

Fig. 4.19 Heme oxidation by oxygen via a μ-peroxo dimer.

through the metal d electrons, which means, of course, that the metal has to be a transition metal. If the metal has no d electrons (e.g. magnesium), or has a full d shell buried within the electronic core of the atom (e.g. zinc), or if there is no metal there at all, then π radical cations or anions are formed.

What is fascinating about the difference between the chemistry of heme on its own, and in the presence of a protein, is the way the latter sterically modulates axial and/or peripheral processes. Consider the reaction of heme with oxygen. Oxygen reversibly binds to the Fe(II) porphyrin via axial ligation. In so doing, it converts high-spin Fe(II) to the low-spin form, which is relatively stable. What brings about oxidation is the next step, *in which there is attack of a second Fe(II) metalloporphyrin on the Fe(II)–oxygen complex to form a* μ-peroxo-bridged dimer. Depending on the ensuing conditions, this bridged complex then collapses to yield Fe(III) metalloporphyrin. Without the attack of the second heme, oxidation would not occur. This kind of behaviour is not unusual in some transition metal elements; cobalt–bridged-oxygen species have been extensively studied in this context.[14]

4.4.2 How proteins alter heme chemistry

If the iron porphyrin is embedded in a suitable protein, then the heme chemistry is completely altered; not least because the protein forms a kinetic barrier to the formation of any kind of oxygen-bridged μ-peroxo-iron–porphyrin dimer. The added subtlety that nature provides is in the exact "cut and style" of the protein coat. We have already examined one style—the wrapping of heme in the protein globin. Here, the protein only allows approach to the heme iron via an axial position. But the attacking axial ligand has to be small enough to wriggle through the steric constraints set up by the globin. Clearly, heme is far too big to fit, but smaller molecules will. Just how hemoglobin and myoglobin can apparently be oxidised by small reducing agents will be dealt with in a later chapter.

It is possible to represent simply the kinetic effects of any particular protein coat around the heme by the letters G and C. The globins are represented by the letter G. This is because the G conformation represents the way globins block the peripheral positions from attack but allow access to the axial position (see Figure 4.20). In the C conformation, the axial positions are blocked but a peripheral position is free to allow redox reactions

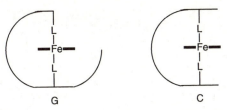

G   C

Fig. 4.20 'G' and 'C' conformations for hemoproteins (e.g. hemoglobin is G; cytochrome *c* is C).

to take place. This is the situation we have in probably the most studied redox-active hemoprotein, cytochrome *c*.

4.4.3 *Heme in cytochrome c*

The cytochromes are probably the most important class of electron-transfer proteins. They are involved in photosynthesis, embedded in the thylakoid membranes of chloroplasts. They are also involved in the opposite process of cellular respiration in plants and animals, which occurs in the membranes of mitochondria.[15]

The heme group in cytochrome *c* is superficially the same as that in hemoglobin. However, a closer look at the molecule will reveal subtle but highly important differences (from the point of view of the way cytochrome *c* functions). For example, the vinyl groups have been converted to thioethers by covalent linkage to cysteine amino acid residues of the cytochrome protein chain. Also, the fifth and sixth coordination sites of iron are occupied by the imidazole nitrogen of a histidine amino acid residue and the sulphur of a methionine amino acid residue. So the heme from cytochrome *c* is called *heme c*. Heme *c* in cytochrome *c* is, thus, firmly covalently anchored to the protein's amino acid sequence, which compares with the looser non-bonding interactions that bind heme to globin in hemoglobin. Apart from providing a different kinetic barrier to the heme than globin (i.e. by blocking the axial instead of the peripheral positions), the way the heme is bound, via the thiother linkages, serves to drastically alter the redox potential of the heme. Also, the heme iron is permanently six-coordinate (bound to four porphyrin nitrogens, an imidazole nitrogen of a histidine residue, and a sulphur from a methionine residue, both in the axial positions) and always low spin in both the Fe(II) and Fe(III) oxidation states. Thus, compared with heme in hemoglobin (with a $E^0 = 0.82$ V), heme *c* in cytochrome *c* has a redox potential of $E^0 = 0.25$V. This means that redox cycling of the heme iron is now far more favourable. In fact, cytochrome *c* is just one of a family of cytochromes that form an electron-transporting chain, sequentially oxidising each other and so passing elec-

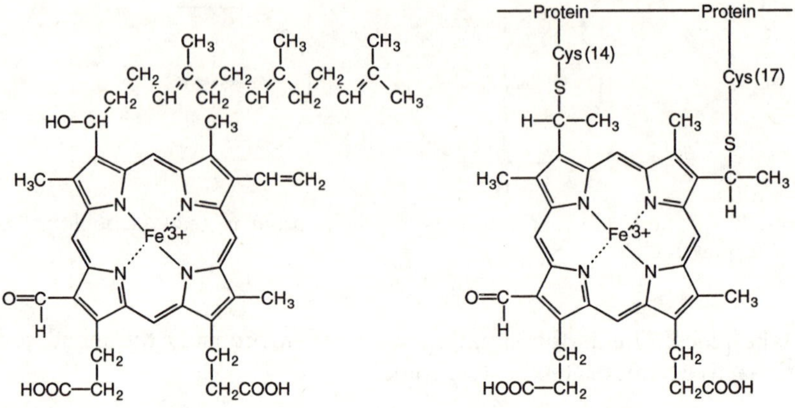

Fig. 4.21 Comparison of heme a (cytochrome a) and heme c (cytochrome c).

trons from one to the next (e.g. b to c_1 to c to $a+a_3$; the suffixes were originally used by their discoverer, David Keilin to note the fact that these cytochromes are distinguished from each other by changes in their light-absorption spectra), terminating with oxygen.

The polypetide chain attached to the heme group in cytochrome c contains a variable number of amino acids; from 103–104 in some fish and earth-bound vertebrates, to 112 in some green plants. The fact that cytochrome c is so central to photosynthesis and respiration indicates that it is probably one of the oldest biomolecules around.[16] It is even possible to create a genealogy of various organisms based on the differences in the amino acid sequences of their cytochrome c. These differences, say between simple yeasts, plants, insects, and ultimately the higher mammals and humans, are not huge, and certainly not enough to alter the way cytochrome c functions. Thus, cytochrome c from one organism will work quite happily in another.

4.4.4 Mechanism of action of cytochrome c

X-ray studies[17] show that cytochrome c is a roughly spherical molecule with a groove down one side. Into this groove fits heme c with one edge (a *meso*-position) slightly protruding. The heme would therefore appear to be ideally placed to undergo fast electron transfer via a peripheral mechanism involving the porphyrin π-system. Although, cytochrome c is probably one of the best studied hemoproteins, it is still not clear exactly what mechanism it uses for the oxidation and reduction of the heme moiety. It seems likely, however, that in the reduced state, the Fe(II) porphyrin undergoes oxidation to form a π-cation radical, which then undergoes rapid internal rearrangement of its electrons so that the Fe(II) becomes oxidised to Fe(III). This would be possible because, in the low-spin state, the t_{2g} set of iron d orbitals are full [Fe(II) has a d^6 electron configuration] and are strongly coupled to the porphyrin π-system. Consequently, any perturbation of the porphyrin π-system, such as oxidation, would immediately be transmitted to the iron atom. This does not preclude the possibility that the heme iron in cytochrome c might be oxidised and reduced via a fast "outer sphere" axial process. Indeed, the available data[18] can be explained using a peripheral or an axial mechanism. It might also be possible for both mechanisms to occur simultaneously, in so far as the oxidation could occur at the peripheral position of the porphyrin, while reduction of Fe(III) back to Fe(II) might involve electron transfer through one of the axial ligands (probably the axial sulphur). This, in turn, would require electrons to be transferred through the protein chain to the heme, so that the sites of oxidation and reduction of cytochrome c could be separate from each other.[19] It is known that the conformation of the protein around the heme is different in the oxidised and reduced forms of cytochrome c.[20] In addition, measurements of the change of E^0 with pH for cytochrome c shows that acidic and basic amino acid residues near the heme are involved in the electron-binding process.

4.4.5 Cytochrome c in cellular electron transport

Electron transport, and its corollary oxidative phosphorylation (the generation of ATP from ADP), are the final and most important stages of cellular respiration. All the enzymatic steps in which high-energy molecules, such as carbohydrates, fats, and amino acids, are oxidised in aerobic cells lead to these last steps: electrons flowing towards oxygen from organic substrates, and reducing the oxygen to water. Energetically, this flow is "downhill" so that electrons are losing energy. This energy is used in oxidative phosphorylation.[21]

It can easily be overlooked just how important ATP synthesis is, for it is the hydrolysis of ATP back to ADP that generates the energy we need to live. For example, an adult male in a sedentary occupation burns about 3000 kcal of energy per day. This requires the daily hydrolysis of about 200 kg of ATP, which presents a problem as the body only has, on average, 50 g of the stuff! So ATP has to be hydrolysed down to ADP and resynthesised back to ATP many thousands of times in the course of the day. *Cytochrome c is just one of the links in the chain of electron carrier proteins that ultimately produce ATP.*

The site of this electron-transporting chain, and the enzymes that produce ATP, are cristae within the cell's mini-power station, the mitochondrion. Each mitochondrion consists of a space, the matrix, which contains

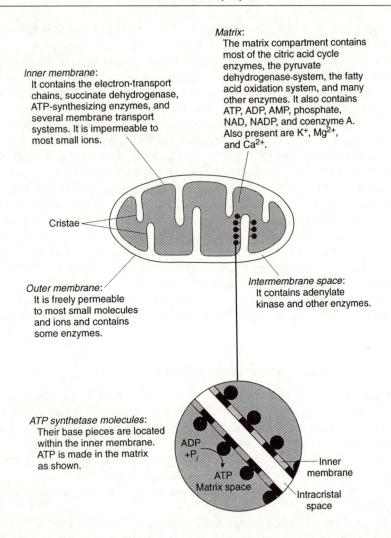

Inner membrane:
It contains the electron-transport chains, succinate dehydrogenase, ATP-synthesizing enzymes, and several membrane transport systems. It is impermeable to most small ions.

Matrix:
The matrix compartment contains most of the citric acid cycle enzymes, the pyruvate dehydrogenase-system, the fatty acid oxidation system, and many other enzymes. It also contains ATP, ADP, AMP, phosphate, NAD, NADP, and coenzyme A. Also present are K^+, Mg^{2+}, and Ca^{2+}.

Cristae

Outer membrane:
It is freely permeable to most small molecules and ions and contains some enzymes.

Intermembrane space:
It contains adenylate kinase and other enzymes.

ATP synthetase molecules:
Their base pieces are located within the inner membrane. ATP is made in the matrix as shown.

ADP +P_i

ATP
Matrix space

Inner membrane

Intracristal space

Fig. 4.22 The biochemical anatomy of mitochondria, showing the location of the enzymes of the citric acid cycle, the electron-transport chains, the enzymes catalysing oxidative phosphorylation, and the internal pool of coenzymes. The inner membrane of a single liver mitochondrion may have over 10 000 sets of electron-transport chains and ATP synthetase molecules. The number of sets is proportional to the area of the inner membrane. Heart mitochondria, which have very profuse cristae and thus a much larger area of inner membrane, contain over three times as many sets of electron-transport systems as liver mitochondria. The internal pool of coenzymes and intermediates is functionally separate from the cytosolic pool.

all the enzymes necessary for the oxidation of molecules necessary for the citric acid cycle. The cristae are finger-like projections of the mitochondrial inner membrane on which the various electron-transporting and ATP-synthesising molecules are lined up. The complete electron-transporting chain consists of around 15 or more chemical groupings in three main sites. These sites are really collections of electron-transporting proteins grouped together as functional complexes.

A fourth complex connects site 1 to site 2. Thus, the first complex in the chain consists of the enzyme NADH dehydrogenase and a group of around five closely linked proteins that contain electron-rich iron–sulphur clusters. These are non-heme proteins vital to the flow of electrons through the chain.[22] The second complex consists of the enzyme succinate dehydrogenase and its collection of iron–sulphur proteins. The third complex contains the hemoproteins cytochromes b and c_1, and one iron–sulphur protein. The final complex, the one that actually reduces oxygen to water, is called cytochrome oxidase and contains two heme groups (heme a and heme a_3) and two copper centres. Ubiquinone, also known as coenzyme Q, is a lipid-soluble quinone which connects the first, second, and third complexes. *Cytochrome c is the electron carrier hemoprotein that links the complexes of site 2 to cytochrome oxidase in site 3.*

The reason why biochemists regard these sites as complexes has to do with the way they are separated from the mitochondrial membrane.[23] But how do they know that these electron-carrier proteins function in the sequence that they do? There are several strands of evidence.

Firstly, there are the standard redox potentials (E^0) of these proteins, which are successively more positive as the chain proceeds towards oxygen (electrons tend to flow from electronegative to electropositive systems, causing a decrease in the available free energy). Second, each member of the electron-transporting chain is specific for a given electron donor and acceptor. So NADH can transfer electrons on to NADH dehydrogenase but cannot short-circuit the chain by transferring them directly to cytochrome b or c. The third strand of evidence is the use of specific inhibitors of the electron-transporting enzymes. Rotenone, a plant toxin used by South American indians to poison fish, stops the flow of electrons between NADH and ubiquinone; the antibiotic, antimycin a, interrupts the flow between ubiquinone and cytochrome c; while cyanide and carbon monoxide block the hemes in cytochrome oxidase. This explains why these two materials are such deadly poisons. They "lock" the heme iron atoms of cytochrome oxidase in the Fe(III) state so that it can no longer function as the reducing agent for oxygen. Consequently, death by cyanide or carbon monoxide means suffocating from the cells upwards.

Interestingly, cyanide and carbon monoxide also bind to the heme of hemoglobin, but the interference of electron transport is much more serious

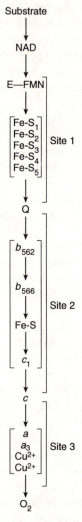

Fig. 4.23 The complete set of electron carriers of the respiratory chain. In site 1, there are at least five different iron–sulphur centres. In site 2, there are two different forms of cytochrome *b*, with different light-absorption peaks, as well as an iron–sulphur centre distinct from those in site 1. In site 3, there are two copper ions in addition to cytochromes *a* and a_3. The precise sequence and function of all the redox centres is not known with certainty.

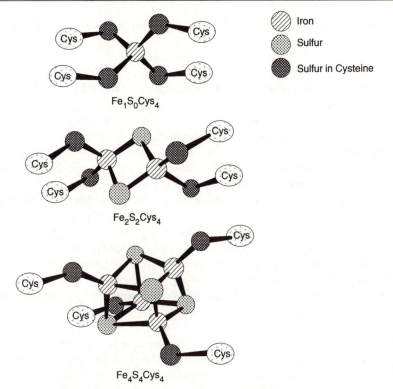

Iron
Sulfur
Sulfur in Cysteine

$Fe_1S_0Cys_4$

$Fe_2S_2Cys_4$

$Fe_4S_4Cys_4$

Fig. 4.24 The structures of iron–sulphur centres.

than oxygen transport. The cure for cyanide poisoning, by inhalation of amyl nitrite, involves oxidation of hemoglobin to methemoglobin, in which the heme iron is oxidised to Fe(III). Methemoglobin binds cyanide much more strongly than heme *a* in cytochrome oxidase and so removes cyanide from the body, even though it is useless for oxygen transport.

4.4.6 *What oxygen does in respiration*

It was mentioned earlier that as the electrons shuttle down the electron-transport chain, their energy is used in the production of ATP via oxidative phosphorylation. An analogy can be made with a waterfall. The water tumbles down a ravine, losing energy as it falls. If a turbine is put in the water's path, the falling deluge will turn the turbine blades, and so useful work is done, i.e. by creating electricity. At the bottom of the fall, the now spent water flows quietly away, eventually out to sea. Following this analogy, during oxidative phosphorylation, the electrons flow downhill, energetically speaking; their energy being used to make ATP as they go. But what happens when all their energy is spent? Which "sea" do the electrons flow into? The answer is oxygen.

Earlier, it was shown that in the molecular orbital description of oxygen, the highest occupied molecular orbitals were only half-filled. In other words, the molecule contains two unpaired electrons and is a biradical. This means that oxygen can quite happily accept more electrons. In so doing, it is reduced, ultimately to water. On the way, however, highly reactive partial reduction products, such as superoxide, $O_2^-\cdot$, peroxide, O_2^{2-}, and the highly lethal hydroxyl radical, $OH\cdot$, can wreak havoc with the sensitive molecules that constitute a cell's structure and function; the lipid membranes of the cell wall, for example. This is the reason why the pollution of the early atmosphere by the photosynthetic activity of the blue-green algae must have led to the disappearance of so many anaerobic bacteria—victims of the corrosive gas oxygen and its partial reduction products—and, because oxygen levels must have been rising rapidly, survival strategies based on avoiding oxygen would have been impractical. The pay-off for becoming aerobic and learning to live with oxygen is, of course, the greater amounts of energy that can be acquired from oxidising carbohydrates to carbon dioxide and water (via ATP synthesis) compared with anaerobic glycolysis.

This is why it is necessary to go to the trouble to ensure that oxygen is reduced quickly and efficiently to water. This is a process that requires a molecule of oxygen to accept four electrons and four protons.

$$O_2 + 4e- + 4H+ = 2H_2O$$

The electrons are provided by the electron-transport chain of proteins, while the protons come from inside the mitochondrion. Cytochrome c plays a vital role in ensuring that electrons are fed to cytochrome oxidase so that it can perform the function of efficient four-electron reduction of oxygen to water.

Cytochrome oxidase has been likened to a "molecular machine". It is present in all animals and plants, aerobic yeasts, and some bacteria. The reaction that cytochrome c oxidase catalyses (Figure 4.25) is the terminal reaction in the sequence of reactions that constitute the respiratory chain. This provides most of the free energy required for life by coupling together electron transport and the synthesis of ATP. Cytochrome oxidase also functions as a proton pump, using some of the free energy released during respiration to pump protons back across the cell membrane against a concentration gradient.

Cytochrome oxidase is a complex protein consisting of two main subunits (labelled I and II) and some smaller ones.[24a] These subunits interlock. The functional unit of this complex protein consists of four redox-active metal centres, two of which are heme a-bound iron atoms of cytochromes a

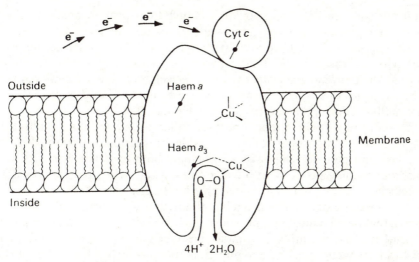

Fig. 4.25 Cytochrome *c* and its relationship to the inner mitochondrial membrane.

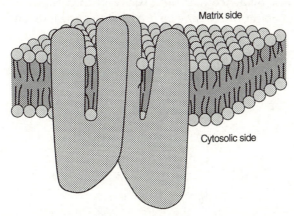

Fig. 4.26 Model of cytochrome *c* oxidase dimer in the mitochondrial inner membrane.

and a_3 in subunit I. The other two redox-active metals are copper ions, Cu^{2+}, one in subunit II and the other also probably in subunit II. Other non-redox-active metals (e.g. Zn and Mg) are probably also involved.

4.4.7 Coping with partial reduction products of oxygen

Clearly, any organism that thrives on oxygen must also have strategies for dealing with superoxide, peroxide, and hydroxyl radicals. The trouble is that all of these substances are interrelated, one begetting the other in aqueous solution.[24b] So, superoxide is known to disproportionate rapidly to hydrogen peroxide and oxygen, while peroxide can react with reduced metal cations, e.g. Fe(II), to give hydroxyl radicals. The latter are the most reactive and can produce longer lived, less reactive intermediates which diffuse to other, more sensitive parts of the cell and do even more damage. The strategies that cells have developed to combat reduced oxygen species are multilevelled and coordinated[25] (see Figure 4.28).

Dismutases (of which there are three different known classes) effect the rapid conversion of superoxide to peroxide. This last is readily removed by catalase and peroxidase enzymes, which are both hemoproteins. Catalases disproportionate peroxide into water and oxygen, while peroxidases use a sacrificial substrate which is regenerated by the cell.

The next line of defence is against the hydroxyl radical, which is so reactive that no enzyme could possibly survive if cells tried to range an attack against it. Instead, antioxidant molecules (such as α-tocopherol and ascorbic acid, or vitamin E and vitamin C as they are better known) probably take care of the hydroxyl radical.

There is now evidence that the toxicity of oxygen, as expressed by its partial reduction products, may be used in the overall defensive strategy of

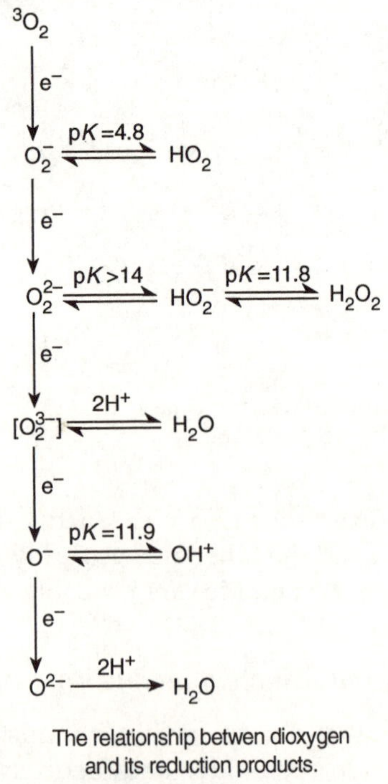

The relationship betwen dioxygen
and its reduction products.

Fig. 4.27 The relationship between dioxygen and its reduction products.

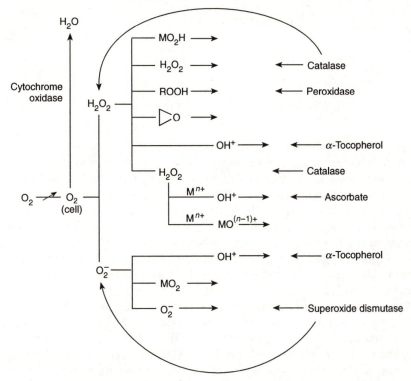

Fig. 4.28 Defensive forces against potentially toxic species derived from dioxygen.

the body.[26] Thus, when white cells attack an invading germ or virus, they begin to consume oxygen. This lasts for about 10 minutes after which digestion continues for some time. All three reduction products, superoxide, peroxide, and hydroxyl radicals, have been observed during this process and are probably instrumental in digesting the pathogen.

Catalases (tetrameric molecules with molecular weights of around 220–250 000 in mammals) and peroxidases (monomeric with molecular weights of 30 000–80 000 depending on the source) are hemoproteins that differ in several important respects from the other hemoproteins discussed so far. Firstly, the iron cycles through a greater range of oxidation states. The native enzymes contain hemin (four in the case of catalase) with the iron in the Fe(III) state. On reaction with hydrogen peroxide, the hemoprotein is doubly oxidised. The iron porphyrin has been variously formulated as being in the Fe(V) state, in the ferryl Fe(IV)=O state, or as a combination of Fe(IV) state and the porphyrin oxidised to a π-cation radical.[27] Secondly, the protein exposes even more of the iron porphyrin to the outside world than either the G or C conformations, in hemoglobin and cytochrome *c*,

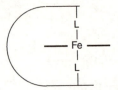

Fig. 4.29 Short 'C' protein conformation in catalases, etc.

respectively. The protein can be thought of as being of a "short C" conformation, which exposes the periphery *and* an axial position to attack. A good
example of the short C conformation is shown by the cytochrome P450 family of hemoprotein enzymes. Their job is to catalyse the addition of an oxygen atom to a substrate, usually a hydrophobic compound such as a drug, a
steroid, or a pesticide. They are also involved in the catabolic breakdown of
hemes in the liver from a variety of sources. They function essentially by
inserting an oxygen atom into an R-H bond to make ROH. The hydroxylation makes compounds more water soluble and so helps in their elimination by the body.

Once again, as with many hemoproteins, the heme moiety is held within
a hydrophobic pocket, but near to the surface of the protein. The short C
conformation, however, allows a hydrocarbon substrate to be bound very
close to the heme (about 5 Å from the iron site). Oxygen is bound axially to
the iron, and the O_2 molecule projects out into the solution surrounding the
protein. The ligand opposite to the bound oxygen is an electron-donating
sulphur atom from a cysteine amino acid residue in the protein sequence.
In its resting state, the iron atom is present as Fe(III) and a water molecule
occupies the sixth position opposite the cysteine sulphur atom. When the
hydrocarbon substrate binds, it transfers an electron to the iron (probably
via a peripheral mechanism) leading to its reduction to Fe(II). Oxygen now
binds to the Fe(II), followed by further addition of an electron from the substrate, and two protons. The overall effect is to produce a ferryl Fe(IV)=O
complex which attacks the substrate by inserting an O atom. The hydroxy-
lated product is then released into the aqueous surroundings and is
replaced by a water molecule. The binding of oxygen at the Fe(II) state has
a competing reaction from carbon monoxide. This leads to an iron complex
that has an absorption at 450 nm and is the reason for the "P450" epithet.

The precise mechanism of oxygen transfer is still not completely clear.
Two possibilities present themselves: the generation of an oxygen radical
species that attacks the R—H bond, and the transfer of an oxygen atom to
the bond. The former mechanism, if it were to happen would be highly
unusual because P450 oxidations are noted for their selectivity and retention of stereochemical and optical configuration of the substrate. Radical
reactions tend to be unselective and non-stereospecific, although if the reac-

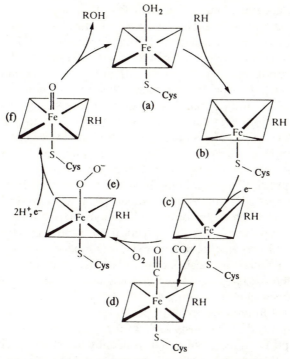

Fig. 4.30 The cycle of reactions of P450. The resulting state of the enzyme is (a) and the important Fe(IV) oxo species is (f).

tion centre experiences tight steric control, it is possible that the selectivity and stereospecificity can be retained.

The oxygen-atom transfer reaction requires that a positive oxygen centre perform an electrophilic attack on the R—H bond. True, this would retain stereochemical configuration (and there is evidence that sterically hindering groups on the substrate that block the approach of an O atom reduce the reactivity of P450), but there is no precedent in the inorganic chemistry of iron for such a reaction. It does, however, point to a special role for the iron-bound sulphur ligand. This has a large polarisability which probably serves to stabilise the high formal charge on the iron atom, so controlling the amount of positive charge that ends up on the transferred oxygen atom.[28]

4.4.8 *An example of a peroxidase enzyme: ligninase*

Lignin is one of the most important natural polymers. Combinations of lignin and cellulose (lignocelluloses) make up as much as 95% of the earth's land-produced biomass (and about one quarter of that is lignin). It strengthens trees and plants and protects them from the weather, insects, and pathogenic organisms. Only a few microbes, such as the white-rot fungus, *Phanerochaete chrysosporium*, have worked out how to penetrate the defensive lignin walls of plants and get to the carbohydrates within.

Lignin is so inert that it contributes to the longevity of biomass and makes plant products like cellulose so difficult to get at. Lignin is also a by-product of the paper-making industry (around 30 million tons are produced every year) and the refining of cellulose and sugar. The structure of lignin contains groupings that could provide a useful source of fine chemicals. Only 5% of this lignin is used at the moment, (a) as a raw material for the food flavouring substance vanillin, (b) as a filler, and (c) as a dispersant and emulsifier. The rest is burnt or dumped, so losing an opportunity to turn a waste product into a potentially valuable source of chemicals.[29]

As biopolymers go, lignin is unusual because it has a completely haphazard structure. In contrast to other well-known biopolymers, such as proteins, DNA, and starch, lignin is produced randomly, by the radical coupling of coniferyl alcohol (4-hydroxy-3-methoxy-cinnamyl alcohol). The lignin polymer therefore consists of many electron-rich aromatic rings that could easily be oxidised. An example of what part of the lignin polymer might look like is shown in Figure 4.32.

The white-rot fungus generates a special lignin-destroying enzyme called ligninase. What had biochemists fooled for a long time was that they could not understand how an enzyme, which usually attacks a specific substrate, could cope with something as random as lignin. They began to think that the white-rot fungus did not deploy an enzyme to attack lignin. Instead, they thought that perhaps the fungus used hydrogen peroxide or hydroxy radicals. This was not such a bad idea, because lignin is degraded at some distance from the developing hyphae of the fungus, which could be taken to mean that the hyphae produce something that diffuses out of the cells, works into the lignin, and then degrades it. Also, hydroxyl radicals are known to attack model compounds of lignin, giving similar compounds to those found in the degradation of lignin. There was also conclusive evidence that hydrogen peroxide was involved in lignin degradation by the white-rot fungus. Professor John Palmer, of London's Imperial College, was one of the first plant biochemists to recognise that ligninase from the white-rot fungus is in fact a peroxidase enzyme.[30]

What happens is that hydrogen peroxide, produced by the fungus's hyphae (from the partial reduction of oxygen) oxidises the ligninase heme to generate the O=Fe(IV) porphyrin cation radical. This then extracts an

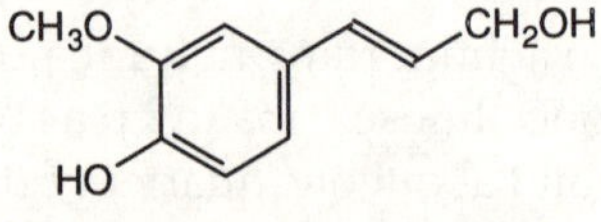

Fig. 4.31 Coniferyl alcohol.

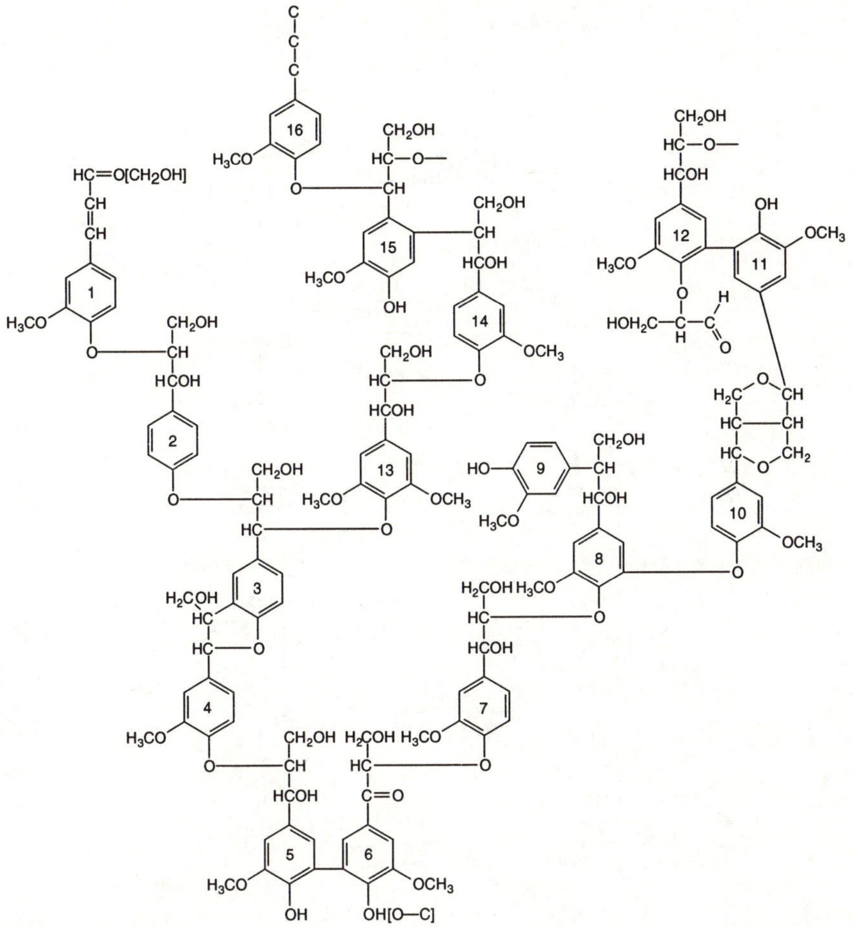

Fig. 4.32 Prominent structures of soft wood lignin comprising 16 phenylpropane units.

electron from one of lignin's electron-rich aromatic groups, producing a more stable aromatic cation radical species. This leads to a cascade of fragmentation reactions as the unpaired electron(s) rampage through the polymer, causing it to degrade gradually, like a zip unzipping.

If ligninase could be mass-produced, then it would be possible to make use of all the lignin that is wasted every year, to generate fine chemicals cheaply. In addition, because of the amount of lignin that ends up in waste water, ligninase could become a novel water purification treatment. In degrading lignin, ligninase is effectively breaking carbon–carbon bonds at ambient temperatures and pressures. Such enzymatic "cracking" could also be used as a low-energy tool for refining petrol.

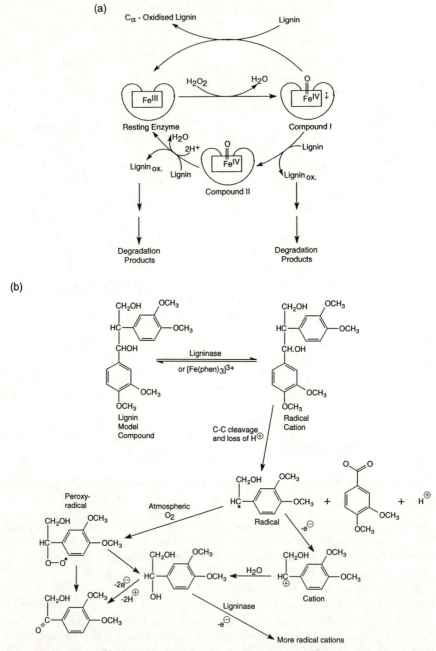

Fig. 4.33 (a) Possible catalytic cycle for the degradation of lignin, assuming lignin-nase behaves as a peroxidase enzyme. [From T.K. Kirk, *Phil Trans. R. Soc. London*, (1987), **A321**, 403.] (b) The degradation of lignin model compound.

4.5 References

1. See F.A. Cotton and G. Wilkinson, *Advanced Inorganic Chemistry*, 5th Edn, John Wiley and Sons, Chichester (1988).
2. See J. Gribbon, *The Hole in the Sky*, Transworld Publishers Ltd, (1988).
3. W. Day, *Genesis on Planet Earth*, 2nd Edn, Yale University Press, New Haven (1984).
4. (a) W.D. Jolly, *Modern Inorganic Chemistry*, 2nd Edn, McGraw–Hill, (1991), p. 107; (b) also ref. 1, p. 106.
5. J.D. Lee, *Concise Inorganic Chemistry*, 4th Edn, Chapman & Hall, (1991), p. 762.
6. D.F. Shriver, P.W. Atkins, and C.H. Langford, *Inorganic Chemistry*, Oxford University Press, Oxford (1990), p. 23.
7. J.E. Huheey, *Inorganic Chemistry*, 34d Edn, Harper International S.I. Edition, (1983), p. 359; also ref. 4(b), p. 390 and ref. 6, p. 205.
8. See J. Baggott, in *Textbook of Biochemistry*, 3rd Edn, ed. T.M. Devlin, John Wiley & Sons, (1992), p. 1025.
9. See R.M. Schultz and M.N. Liebman in ref. 8, p. 119.
10. See R.G. Shulman, S. Ogawa, K. Wuthrich, T. Yamane, J. Peisach, and W.E. Blumberg, *Science* (1969), **165**, 251.
11. See M. Perutz, *Nature*, (1970), **228**, 726; *Sci. Am.*, (1978), **239**, 92.
12. See C.E. Castro in *The Porphyrins*, Vol. 5, ed. D. Dolphin, Academic Press, New York (1978), p. 1.
13. See ref. 4(b), p. 494; ref. 6, p. 486; ref. 7, p. 559.
14. See ref. 1, p. 500.
15. See M.S. Olson in ref. 8, p. 276.
16. See S. Ferguson-Miller, D.L. Brautigan, and E. Margoliash, in *The Porphyrins*, Vol. 7, ed. D. Dolphin, Academic Press, New York (1978), p. 149.
17. See R. Timkovich in ref. 16, p. 241.
18. See ref. 6, p. 622.
19. See ref. 12, p. 19.
20. T. Takano, *J. Biol. Chem.*, (1977), **252**, 776.
21. A.L. Lehninger, *Principles of Biochemistry*, Worth, (1982), p. 467.
22. See ref. 7, p. 865.
23. See ref. 21, p. 484.
24. (a) B.G. Malmstrîm, *Chem. Rev.*, (1990), **90**, 1247; (b) D.T. Sawyer, *Oxygen Chemistry*, Oxford University Press, Oxford (1991).
25. C. Greenwood and H.A.O. Hill, *Chem. Brit.*, (1982), **18**, 194.
26. J.M. McCord, D.K. English, and W.F. Petrone, in *Biological and Clinical Aspects of Superoxide and Superoxide Dismutase* eds W. H. Bannister and J. V. Bannister, Elsevier, New York (1980), p. 154.
27. See W.D. Hewson and L.P. Hager, in ref. 16, p. 295.
28. See B. Walker-Griffin, J.A. Petersen, and R.W. Estabrook in ref. 16, p. 333.
29. D.W. Ribbons, in *Technology in the 1990's: Utilisation of Lignocellulosic Wastes*, Phil. Trans. R. Soc. London, (1987), **A 321**, 403.
30. See ref. 29, p. 495.

5. ... And where porphyrins go to

5.1 Introduction

The great science fiction writer Arthur C. Clarke once wrote that behind every living human being stands a hundred million ghosts. It is another way of saying that living things die and replace themselves. But it's not only living things (right down to the lowliest virus) that turn themselves over. Many of the molecules that control biological activity do the same thing and porphyrins are a good example of this. Red blood cells live for about three months in the bloodstream, carrying oxygen around the body, before becoming senescent. They are then transported to the liver where they are dismantled. During this process, the heme is broken open to release the iron, which is either excreted or saved depending on the body's reserves of the metal. After several further transformations, the remains of the heme are excreted from the body.

Chlorophyll in plants is also continually turned over, and its creation and destruction seems to depend on the amount of light available. Try camping in a field for a week and then leave. The grass where your tent has been standing will be yellow from lack of chlorophyll, compared with the grass that has been continually exposed to the ambient light.

Living systems know how to synthesise porphyrins. They also know how to break them down. What are the mechanisms of this breakdown, and what are the compounds they are broken down into? We shall see how such a question becomes, at least as far as the animal kingdom is concerned, increasingly scatological.

5.2 Bile pigments

5.2.1 Occurrence

Animals break down their heme into compounds called bile pigments. These are open-chain tetrapyrroles, such as biliverdin and bilirubin, whose names were acquired when they were first separated from animal bile.[1] As these names further imply, biliverdin is green, while bilirubin is a yellowish orange. Many other open-chain tetrapyrroles that are separated from natural sources, not only have no connection with bile, but some of them are even colourless. Nevertheless, in their names these compounds still

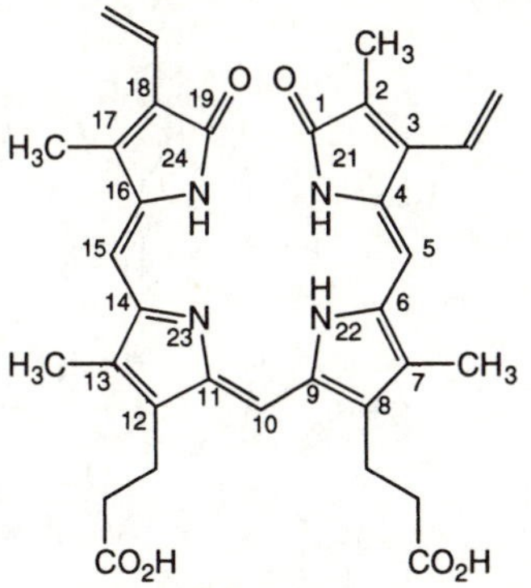

Fig. 5.1 Comparison of biliverdin and bilirubin structures.

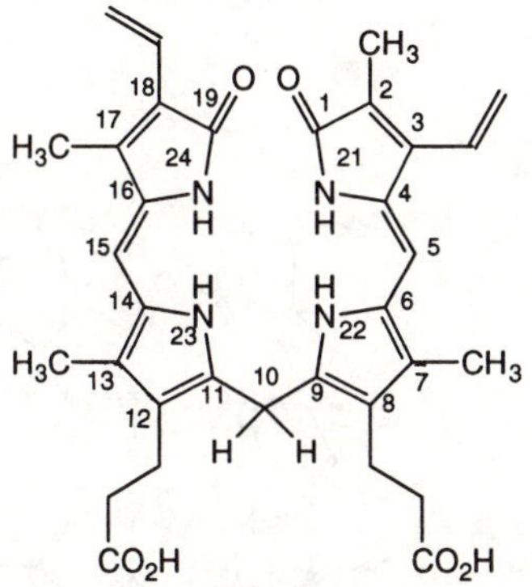

retain the animal bile connection (see Chapter 2 for the nomenclature of compounds of the bile pigment family).

Biliverdin crops up in many niches in the animal kingdom; for example, as the green skin pigment in certain reptiles and amphibians. Some insects are green because of biliverdin, e.g. the skin of the caterpillar of the cabbage white butterfly. Many birds will also lay down biliverdin in their egg shells.[2] In the animal kingdom open-chain tetrapyrroles (termed *bilins* or *bilinoids*) are usually a product of excretion, so that even though the bilin might be serving a useful purpose (e.g. camouflage in a green environ-ment), this is of secondary importance compared to the primary need to excrete these metabolites from the system.

In the world of photosynthetic organisms, it is a totally different story. Bilins, though catabolically derived from heme or chlorophyll, are of primary importance as photoreceptors and photosynthetic light harvesters. Thus, phycocyanobilin and phycoerythrobilin, suitably attached to polypep-

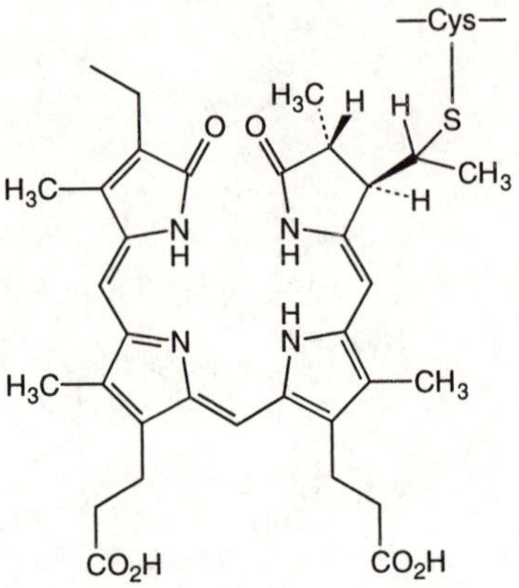

Cys-Phycocyanobilin

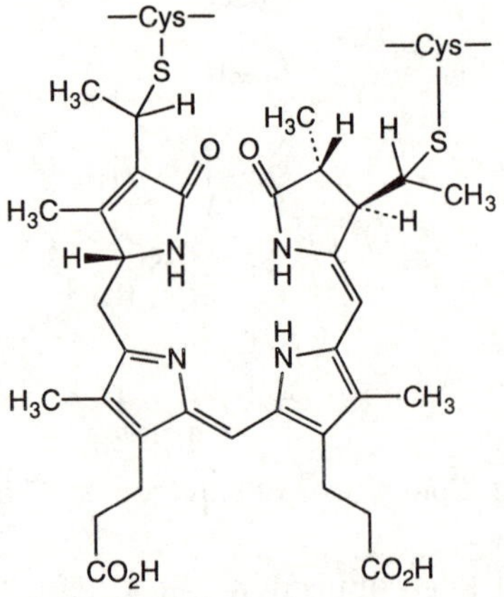

DiCys-Phycoerythrobilin

Fig. 5.2 Phycocyanobilin and phycoerythrobilin—plant 'bile' pigments.

tide chains, act as light receptors in algae. Related bilins act as the prosthetic groups for a class of plant photoreceptors called phytochromes, which govern many of a plant's responses that are triggered by light; for example, germination and flowering.[3] Consequently, in plants, open-chain tetra-pyrroles are biosynthetic targets, as well as being catabolic end-products.

5.2.2 *Formation of bile pigments requires oxidative ring-opening*

The formation of a bilin from an iron porphyrin not only involves opening the ring, but also loss of the *meso*-carbon at the position of opening, usually pyrrolenine rings withdrawing electron density from *meso* carbons

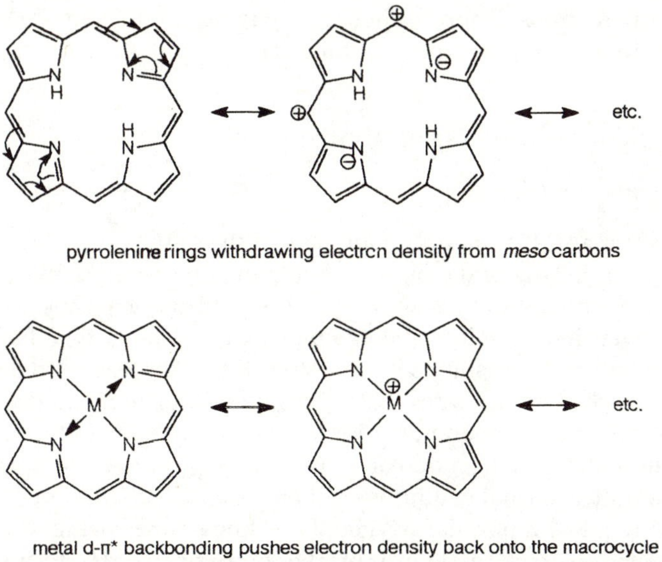

metal d-π* backbonding pushes electron density back onto the macrocycle

Fig. 5.3 Metals push electron density back on to the macrocycle.

in the form of carbon monoxide. Interestingly, without the iron the porphyrin is far less sensitive to the kind of oxidative attack at the *meso*-position that generates bilins. Also, it has been demonstrated that when phycocyanobilin is generated by blue-green algae, they too produce carbon monoxide, a fact that strongly implicates heme as the precursor for phyco-cyanobilin formation.[4]

One of the reasons why this oxidative ring-opening of hemes is so interesting is because the *meso*-positions are usually thought of as electron deficient. As was discussed earlier in Chapter 3, the pyrrolenine-type rings may be thought of as withdrawing electron density from the *meso*-positions in order to make up their aromatic sextet. A transition metal, such as iron, has d orbitals that interact synergically with the porphyrin π*-system, pushing electron density back on to the *meso*-carbons. It is therefore easier for oxidation to occur at these positions in heme than in the metal-free porphyrin. But there is another interesting feature about this reaction. All porphyrins have four *meso*-carbons and in a molecule as unsymmetrical as protoheme, each of these *meso*-carbons is different. The oxidative cleavage of heme to give biliverdin (and its subsequent reduction in mammals to give bilirubin) should yield, therefore, four different bilins. If we label the *meso*-carbon at position 5, α, that at position 10, β, position 15, γ, and position 20, δ, then we should have a mixture of bilirubins IX_α, IX_β, IX_γ, and IX_δ. In fact, when heme on its own undergoes oxidative cleavage such a mixture is obtained. *In vivo*, however, the majority of bilirubin obtained is the IX_α

isomer (there are exceptions: thus, the biliverdin found in the skin of cabbage white caterpillars is the IX_γ isomer[5]).

5.3 Mechanism of heme catabolism

5.3.1 Heme to biliverdin

In adult humans, the catabolism of heme is the only known source of bilirubin $IX\alpha$, 300–500 mg of the pigment being manufactured daily. It is a puzzle just why mammals should go to the trouble of reducing the water-soluble green biliverdin to insoluble yellow bilirubin. The body then has to esterify bilirubin with various sugars in order to solubilise it ready for excretion, via the bile duct, into the intestinal tract. Here, bilirubin undergoes further transformation to urobilinoids prior to its final excretion in feces.

Bilirubin and carbon monoxide are always produced in equimolar proportions under normal conditions of heme catabolism; the process is oxidative and requires molecular oxygen. Labelling studies using $^{18}O_2$ show that the labelled oxygen ends up in the bilirubin at the terminal lactam positions, and in the carbon monoxide that is liberated (by removal of the C-5 carbon).[6] The overall process is shown in Figure 5.5.

Most of the bilirubin produced (about 80%) comes from the catabolism of heme from hemoglobin, during the degradation of erythrocytes (red blood cells). The rest comes from the turnover of other hemoproteins, such as cytochromes P450 and b_5, which have a high abundance in, and are rapidly metabolised by, the liver. Proof that not all bile pigments originate from the destruction of hemoglobin (but also arise from the turnover of other hemoproteins) comes from isotopic labelling experiments. Thus, human volunteers fed with ^{15}N-labelled glycine, showed the presence of labelled urobilins in their feces long before newly synthesised red blood cells had been degraded.[7]

Much of what is known about heme catabolism comes from *in vitro* studies using heme, oxygen, and the reducing agent ascorbic acid (vitamin C). The initial stage of heme catabolism is a controlled coupled oxidation catalysed by an enzyme, heme oxygenase, which uses molecular oxygen (as an oxidising agent) and electrons from NADPH provided by a reductase enzyme.[8] A likely early intermediate in this process is a 5-oxyhemin [in which the central metal has also been oxidised to Fe(III)], which may undergo electron transfer from the porphyrin π system to the metal, reducing the latter to Fe(II) and generating a porphyrin π-cation radical. The latter probably reacts further with molecular oxygen to generate a hydroperoxy radical which collapses, with the expulsion of carbon monoxide, to Fe(II) verdoheme. This would react further with oxygen in the presence of a reducing agent (probably NADPH), to generate the Fe(III) complex of biliverdin which, on hydrolysis, rapidly loses iron. The liberated iron is taken up by the iron-storage protein, ferritin.

Why ring-opening should occur selectively at the α-carbon of heme, and not at any of the *meso*-positions, is simply answered by comparing the products from heme cleavage on its own (i.e. minus its apoprotein) and heme in association with a protein, e.g. myoglobin or hemoglobin. Coupled oxidation of free heme gives almost equal amounts of the four possible biliverdins. Heme in myoglobin undergoes coupled oxidation virtually exclusively to biliverdin IXα. Clearly, the apoprotein is providing steric hindrance to the oxidation reaction at all the *meso*-carbons of heme except the C-5 position.[9]

What is the nature of the enzyme that catalyses heme catabolism? As long ago as 1935 Lemberg[10] questioned whether an enzyme was necessary at all, proposing that "the heme iron worked on its own molecule." Although, under the circumstances, this was a highly insightful if unusual proposal, the evidence at that time, pointed to the involvement of an enzyme with activity similar to cytochrome P450. This enzyme was called *heme oxygenase*. Thus, the initial stage of heme catabolism (i.e. hydroxylation of the heme molecules) closely resembles the way cytochromes P450 biotransform drugs and steroids. However, there is now a considerable body of evidence against the notion that heme oxygenase is a cytochrome P450.

For example, the distribution of heme oxygenase around the body is nothing like that of cytochrome P450, the highest specific activity of the former being found in microsomes of the spleen, followed by those of the liver.[11] Next, there are the compounds that inhibit and activate both enzymes. In each case, they are totally different. Heme oxygenase is inhibited by carbon monoxide, while the usual P450 inhibitors have no effect on it. Similarly, P450 activators do not stimulate heme oxygenase activity. On top of this, there is no positive correlation between heme oxygenase activity and the level of P450 in any microsomal preparation—if anything, there seems to be an inverse relationship between them.

Perhaps the most convincing piece of evidence that the two enzymes are different is that purified heme oxygenase (from the pig spleen) contains no detectable cytochrome P450. So, although the hydroxylating activity of the two enzymes is similar, heme itself is the substrate for heme oxygenase.

It appears that heme oxygenase is embedded in the membrane of the endoplasmic reticulum in such a way that when heme is bound (into a crevice of the enzyme), it is in close proximity to the microsomal electron-transport system that also drives cytochromes P450. It is possible that the aproprotein of P450 is related in some way to heme oxygenase and it is interesting to note that hemoglobin, myoglobin, and P450 all have globin-type apoproteins and easily removable heme moieties.[12]

The enzyme-bound heme has a ligand attached to it that provides a strong ligand field. This means that the Fe(III) cation (which has a $3d^5$ outer electron configuration) is in a low-spin state, with the majority of its 3d electrons spin-paired. In this configuration, the Fe(III) cation is easily reduced to low-spin Fe(II) by a suitable electron donor (such as NADPH). Ready

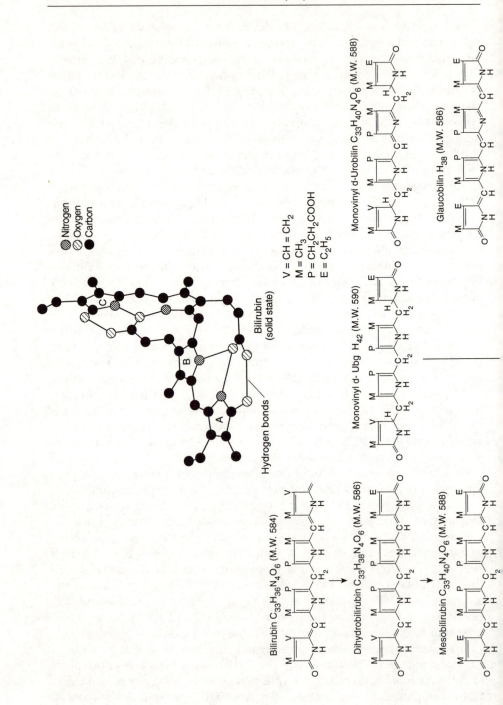

Nitrogen
Oxygen
Carbon

Bilirubin
(solid state)

Hydrogen bonds

$V = CH = CH_2$
$M = CH_3$
$P = CH_2CH_2COOH$
$E = C_2H_5$

Bilirubin $C_{33}H_{36}N_4O_6$ (M.W. 584)

Dihydrobilirubin $C_{33}H_{38}N_4O_6$ (M.W. 586)

Mesobilirubin $C_{33}H_{40}N_4O_6$ (M.W. 588)

Monovinyl d- Ubg H_{42} (M.W. 590)

Monovinyl d-Urobilin $C_{33}H_{40}N_4O_6$ (M.W. 588)

Glaucobilin H_{38} (M.W. 586)

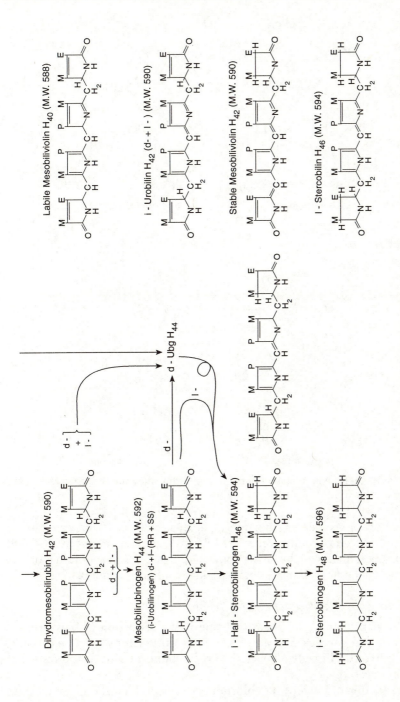

Fig. 5.4 Interrelationships of bile pigments with special reference to biogenesis of the urobilinoids from bilirubin. The vertical sequence of formulae on the left, with the alternative pathways indicated by the horizontal and curved arrows, are believed to represent the principal biogenetic pathways as indicated by present observations. The formulae shown on the right and that of *l*-half-stercobilin in the lower middle are those of oxidation products of the compounds of biogenetic significance at the left.

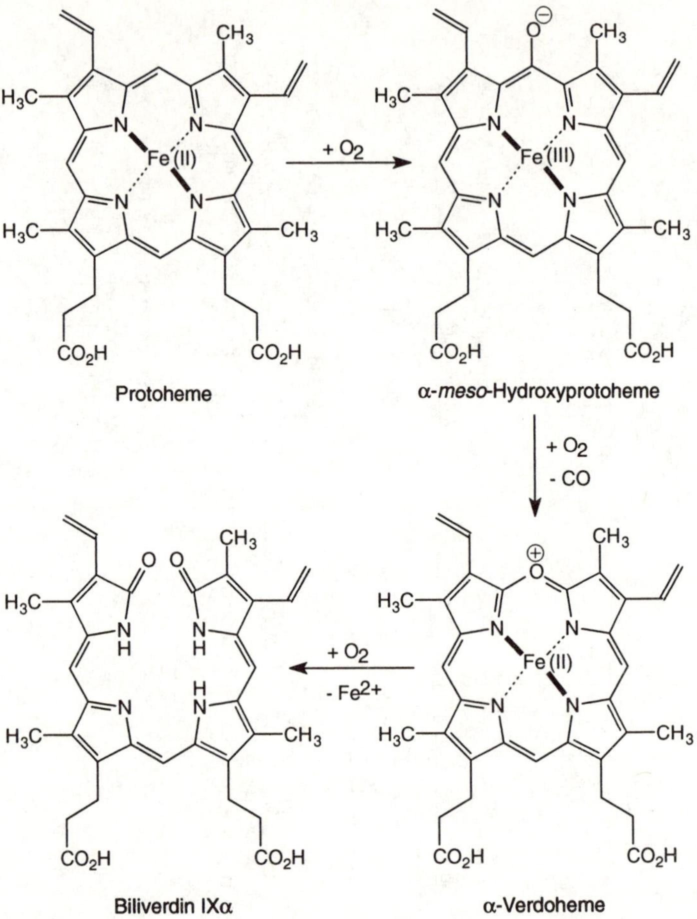

Fig. 5.5 Pathway for conversion of heme to biliverdin, showing the structures of the principal intermediates.

coordination of oxygen occurs, which is then reduced. This generates a radical (a reduced oxygen species like OH·) which then attacks the heme at the α-position to form the α-hydroxylated hemin. Thus, binding of the heme into the heme oxygenase catalyses its own hydroxylation.

5.3.2 Biliverdin to the excretion of bilirubin

The next step in the catabolism of heme is the reduction of biliverdin to bilirubin. Birds, reptiles, and amphibians are quite happy to excrete water-soluble biliverdin directly—hence the green colour of bird droppings—but not mammals or fish.

The reduction of biliverdin to bilirubin creates problems. Bilirubin is both cytotoxic and lipid-soluble, so that an excess build-up in the system (called jaundice) can be fatal. Mammals have solved the problem of bilirubin build-up by generating a specific, enzymatic system for diesterifying bilirubin with glucuronic acid. This polar ester group makes bilirubin soluble in more aqueous media. The enzyme that performs the reduction, called biliverdin reductase, specifically reduces the IXα isomer. The other isomers of biliverdin are either not reduced or reduced at much slower rates. The reductase enzyme is found mainly in the spleen, kidney, and liver, where its activity far outstrips that of heme oxygenase. Consequently, there should be little or no biliverdin in mammalian bile—any that does appear is due to autoxidation of bilirubin in the bile duct.[13]

After the reduction of biliverdin, the bilirubin that is formed then undergoes a series of transport and transformation steps which ultimately lead to its excretion in the intestinal tract. From the sites of its production, bilirubin is released into the plasma where it efficiently binds to albumin, which acts as a plasma-transport system. The bilirubin–albumin complex is carried in the plasma to liver cells (hepatocytes), where the bilirubin is released from its albumin carrier protein and transported across the cell bilayer membrane into the hepatocyte. Once inside, the bilirubin is bound in the cytoplasm to anion-binding proteins such as ligandin. The latter carries the bilirubin to membrane-bound enzymes (localised in the endoplasmic reticulum of the liver cell) which catalyse the esterification of bilirubin, the ester groups (mainly β-D-glucuronoside, but also smaller amounts of β-D-xylopyranosides and β-D-glucopyranosides) being transferred from their uridine diphosphate nucleotides.

The esterification step is essential for bilirubin excretion, not only because it renders the bilirubin more soluble in aqueous fractions, but also because it is now less soluble in lipid membranes. This means the bilirubin cannot flow back into cells, e.g. across the intestine back into the blood stream or across the placenta or the blood–brain barrier. The esterified bilirubin is then collected and excreted into the bile against a large concentration gradient. Once in the intestine, the excreted bilirubin undergoes a series of stepwise reductions, catalysed by the enzymes of intestinal flora, which convert it into urobilinoids (see Figure 5.4). At some point, which is not yet clear, the esterifying groups are also cleaved off. We shall see, in the next chapter, what happens when bilirubin levels are too high and how this condition is dealt with in newborn babies.

5.4 Bile pigments from plants

As plants do not have bile ducts (or anything resembling an intestinal tract, even in carnivorous plants such as the venus flytrap and sundew), the term "plant bile pigment" is a misnomer. Nevertheless, open-chain tetrapyrroles or bilins are more important for plant and algal biochemistry than animal biochemistry, because they function in many of the photobiological responses of these organisms. For example, plant bile pigments act as photoreceptors in photosynthesis (the phycobiliproteins of blue-green, red, and cryptomonad algae) and as photoactive hormones (the phytochromes), which, in trace quantities, control light-dependent changes, e.g. growth and development of plants (not algae). In addition, chlorophyll is continually turned over in plants: the golden brown colours of autumn in temperature climates bear testimony to the mass destruction of chlorophyll at the end of the growing season. Recently, a bilin has been separated from chlorophyll breakdown products.

Phytochrome and phycobiliproteins consist of apoproteins to which open-chain tetrapyrroles are bound as chromophores (see Figure 5.2). As with animal bile pigments, there is strong evidence to suggest that formation of these chromophores depends on oxidative cleavage of heme to give biliverdin. The difference in algae is that the enzyme that accomplishes the oxidative ring-opening of the heme (the algal equivalent of heme oxygenase) is water-soluble—in animal cells, heme oxygenase is insoluble in water—and the redox transaction is mediated by different enzymes.[14] Animal cells utilise cytochrome *c* and NADPH, whereas algae use the non-heme redox enzyme, ferrodoxin. The binding of open-chain tetrapyrrolic chromophores to the apoprotein is usually via thioether links to the vinyl groups on rings A, D, or both. The peptide sequences around the points of attachment of these pigments are distinctive for each type of linkage.[15]

The function of the phycobiliproteins in algae is to harvest light energy. In this respect, they are the algal equivalents of the accessory pigments in plants, acting as antennae, which collect (shorter wavelength) light energy outside the absorption range of the main chlorophyll antenna system. As accessory pigments, phycobiliproteins are highly efficient at transferring light energy to chlorophyll *a*, about 90% of the energy being transferred by a mechanism called inductive resonance. Energy is absorbed by the phycobiliprotein chromophore and, as the excited state collapses, the excitation energy passes to the chlorophyll antenna system. The collection is efficient enough for hardly any of the transferred energy to be lost via fluorescence from the bile pigment chromophore. In other words, the fluorescence of the bile pigment in the natural state (i.e. within a biomembrane) is quenched. Mutant organisms have been raised which lack chlorophyll and whose bile pigments fluoresce when excited by light.

The conditions for efficient energy transfer by inductive resonance are that the phycobiliprotein chromophore (the donor) and the chlorophyll antenna molecule (the acceptor) be in close proximity, usually about 5 nm apart. The "resonance" part is how the frequency of the flurorescence emission spectrum of the donor matches the absorption spectrum of the acceptor. For ideal resonance, the two spectra should overlap as much as possible.

Phytochrome undergoes a photoreversible photoreaction which switches the protein between two states. This simple switching action is believed to be the ultimate trigger for a multitude of responses within the plant which regulate, for example, its growth and control the times of flowering and germination of seeds; all making for difficulty in rationalising a unified reaction scheme for so many complex operations from such a simple response. What is known is that the two states have different absorbtion maxima, which are probably related to conformational changes in the phytochromes' open-chain tetrapyrrole.

It is interesting that there are significant differences in the spectroscopic properties of these bile pigments when they are attached to their native proteins and when they are in a free, uncomplexed state. The free pigments are known to adopt a helical conformation, whereas in the complexed state X-ray crystallography shows that the tetrapyrroles are stretched out. The light reaction that is thought to occur, especially in phytochromes is a *cis–trans* isomerisation about the central double bond,[8] similar to that which occurs between *cis-* and *trans*-stillbene. Some confirmation for this proposal has been obtained by incorporation of a synthetic open-chain tetrapyrrole into a cyclophane.[16] When this blue compound is irradiated with light, isomerisation occurs about the central double bond from *cis* to *trans* to give a new cyclophane with a stretched out tetrapyrrole and a magenta colour. This corresponds to changes in the UV–visible spectrum of the two compounds in which the extinction coefficient of the stretched tetrapyrrole in the visible region is larger, while in the UV region it is smaller, than the helical tetrapyrrole. Similar changes are observed in the native biliproteins.

Every year, in temperate climates, as summer ends so leaves turn brown and drop from the trees. This represents an annual catabolic turnover of chlorophyll of over a thousand million tons, and yet not much is known about how this process occurs. Recently, however, a plant metabolite was separated from the leaves of a mutant strain of barley, that has the structure of an open-chain tetrapyrrole which is clearly derived from chlorophyll.[17] As can be seen in Figure 5.7, the chlorophyll macrocycle has been ring-opened at the C-5 position, which resembles the oxidative ring-opening of heme. Are there other plant bile pigments that are structurally derived from chlorophyll? The most recent answer is yes: the light-emitting substance of krill (part of sea plankton) and luciferin, from dinoflagellates, have been identified as bile pigments derived from chlorophyll. Here, however, the ring-opening has occurred at the C-20 carbon, which in chlorophylls is the *meso*-carbon most susceptible to attack by electrophilic reagents.

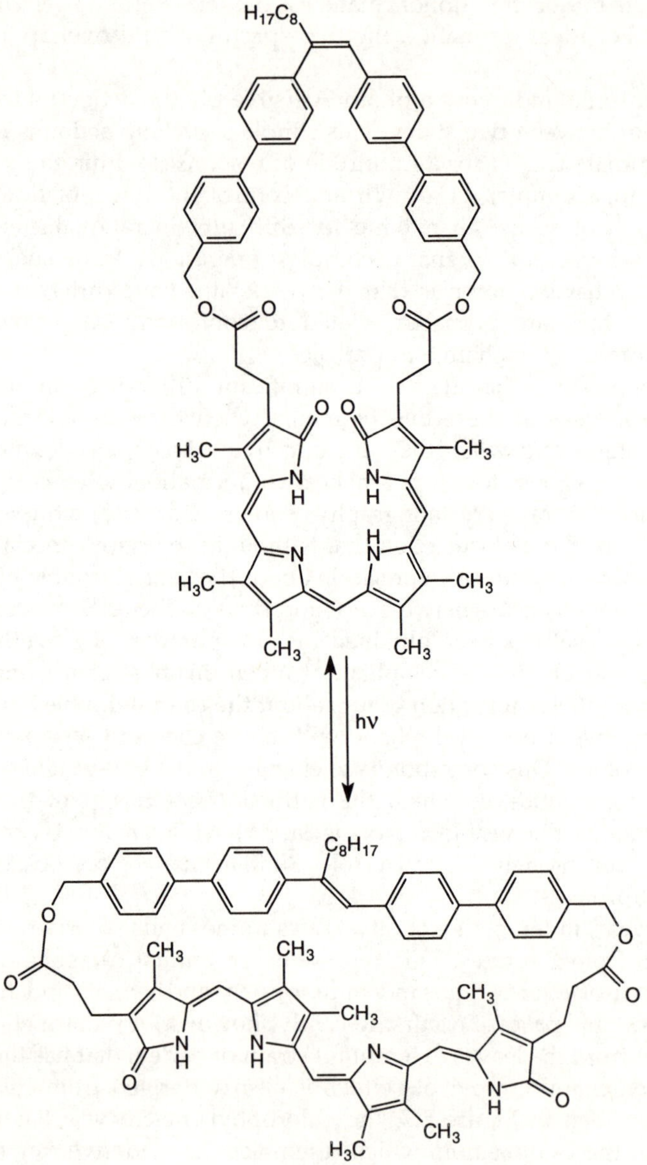

Fig. 5.6 Model system demonstrating light-induced *cis–trans* isomerism in phytochromes.

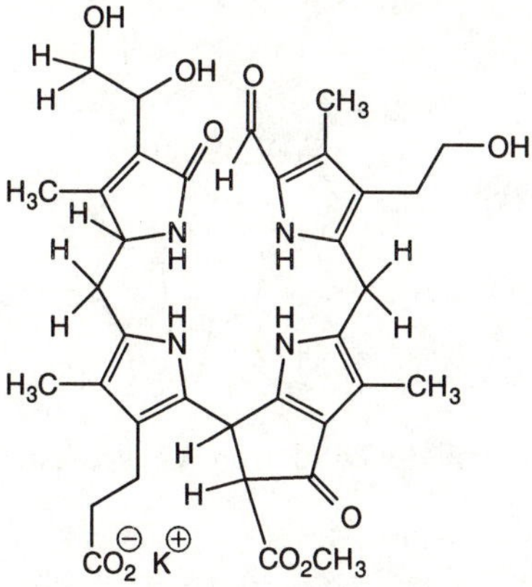

Fig. 5.7 Secoporphinoid catabolite from degradation of chlorophyll.

5.5 Porphyrins from the past

The plant bile pigments and chlorophyll degradation products considered so far are a feature of biochemical reactions occurring in the present. A second kind of chlorophyll degradation product is obtained from geochemical reactions that have gone on for millions of years deep in the earth's crust. When primeval forests and other photosynthetic organisms died long ago, they were buried and subjected to various degrees of chemical and thermal stress (collectively called diagenesis) deep within the earth. Over the eons, this gradually changed their chlorophyll and bacteriochlorophyll into porphyrins that, in small but significant quantities, are associated with coal, oil, and shale deposits. To the geologist looking for new reserves of non-renewable energy, these *petroporphyrins*, act as markers for the type of coal or oil they are associated with and how difficult such deposits are to extract from the ground.[18]

The major metalloporphyrin in petroleum and oil shale is *vanadyl deoxophylloerythroetioporphyrin* (VODPEP) which was first discovered by Alfred Treibs (of the same Munich school of porphyrin chemistry as Hans Fischer) in 1934. Treibs quickly realised that his discovery strongly implied a biological origin for coal and oil. But he went further and developed a scheme of chemical (as opposed to biochemical) reactions that could have occurred to convert chlorophyll in green plants into VODPEP.

Before discussing Treib's scheme, it is worth remembering that in his day, there were none of the sophisticated spectroscopic tools that chemists take for granted these days, such as UV–visible and mass spectroscopy. All he had was a hand-held spectroscope which required a highly experienced eye to use, and the (then) laborious chore of chemical analysis. The Munich

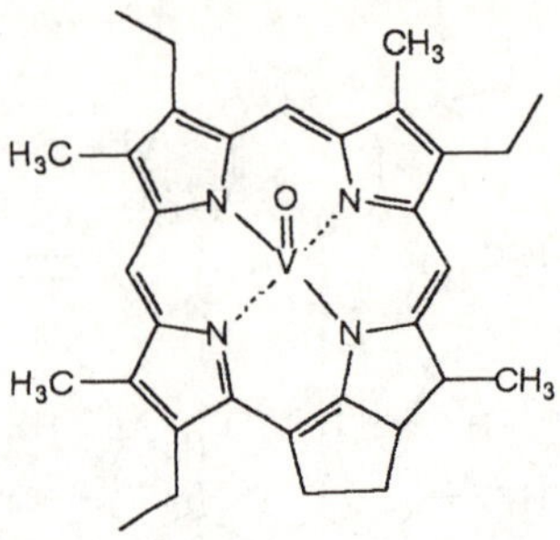

Fig. 5.8 Vanadyl DPEP from Venezuelan crude oil.

school developed both these techniques to a fine art, enabling them to assign the structures of many porphyrins. One of these was VODPEP.

Since Treibs' time, modern analytical techniques have unearthed a vast number of petroporphyrins, in the light of which Treibs' original scheme has had to be modified. Nevertheless, the basic principles that he enunciated still hold, so that a knowledge of the main petroporphyrin types in any particular sample of coal or oil, and their relative concentrations, can provide geologists with accurate information about the thermal stresses and sedimentological history of the environment the material has been exposed to. This, in turn, gives geologists vital clues about the difficulties likely to be encountered in obtaining non-renewable resources.

The identification of petroporphyrins rests heavily on UV–visible and mass spectroscopy. In the former, the region of the UV–visible spectrum that gives the most information are the Q bands, between 480–700 nm. As was explained in Chapter 3, the Q band structure is a powerful indicator of the substitution pattern around the porphyrin macrocyclic nucleus. The pattern of Q band intensities that goes IV>I>II>III, is indicative of a phyllo-type substitution pattern, typically displayed by DPEP. As a vanadyl complex, the Q band structure collapses to two bands. So, once a pure sample of VODPEP has been obtained, mass spectroscopy gives an accurate molecular mass. Acid work-up, followed by neutralisation, generates a metal-free porphyrin, whose UV–visible spectrum shows the typical DPEP Q band structure, while the mass spectrum confirms the loss of the V=O unit.

However, mass spectroscopy of petroporphyrins is usually a much more complicated business than that just described, which dealt only with pure compounds. Any geological sample of coal, oil, or shale, will contain a large number of porphyrins and mass spectroscopy is usually performed on the mixture of porphyrins after they have been separated from their carbonaceous, oily, or bituminous residue. Using this technique, several homologous series have been discovered, the major ones based on DPEP (giving

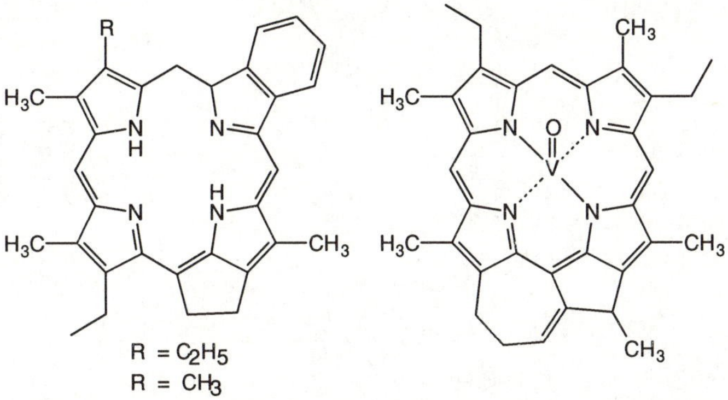

R = C₂H₅
R = CH₃

Fig. 5.9 Some exotic petroporphyrins recently isolated from oils shales.

mass ions with molecular weights of 308 plus increments of 14) and etio-porphyrins (in which tougher geochemical reaction conditions have broken open DPEP's isocyclic ring to yield alkylporphyrins, giving mass ions with molecular weights of 310 plus increments of 14). Other minor series have been discovered, not all of them isolated and identified. In addition, other metals have been discovered complexed to the porphyrins, notably nickel. In the last few years some quite exotic petroporphyrins have been identified.[19]

We shall not deal with the detailed statistical analysis of petroporphyrin mass spectra, the interested reader being referred to more authoritative texts.[18] It is enough to say that the ratio of DPEP-type to etio-type porphyrins in a sediment is a good indicator of the amount of thermal stress that the sediment has been subjected to: the higher the ratio of etioporphyrins, the greater the degree of thermal stress.

Returning to the Treibs scheme, he postulated about six or seven reactions, the order of which were reckoned to be interchangeable, depending on the environmental conditions. Thus, chlorophyll *a* undergoes magnesium loss, followed by removal of the phytyl and methyl ester groups. Next, reduction converts the vinyl to an ethyl group, while also removing the exocyclic carbonyl group. A most important step is the aromatisation of ring D via dehydrogenation of the chlorin nucleus to a porphyrin. This is followed by facile decarboxylation which removes the COOH group on the exocyclic ring, and, at a more leisurely pace, the propionic acid group. Finally, chelation with V=O or nickel occurs.

Treibs thought that the order of these reactions was determined by increasing thermal stress, so that they may not necessarily occur in the order given. The essential point about the scheme, however, is that only one product, VODPEP, is generated from the diagenesis of chlorophyll *a*. Treibs also thought that these reactions would generate a similar end-product if the starting compound was chlorophyll *b*, *c*, *d*, or bacteriochlorophyll.

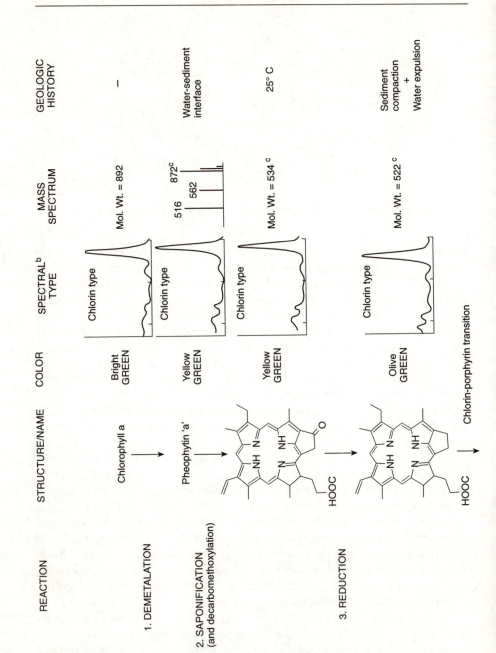

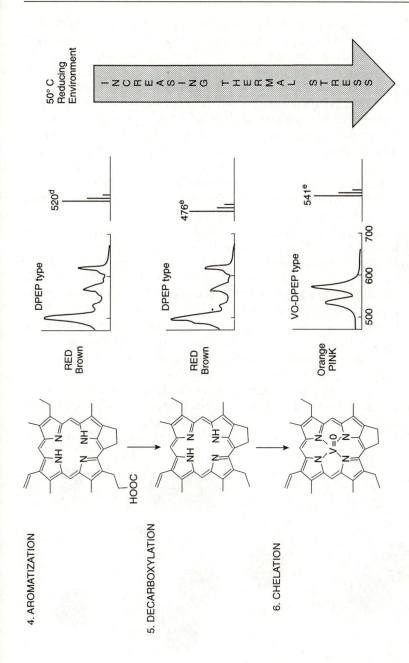

Fig. 5.10 Treib's scheme for chlorophyll degradation to petroporphyrins.

Since Treibs' time, a vast number of petroporphyrins have been discovered, so that it has been necessary for Treibs' scheme to undergo substantial modification in order to include other reactions that could even begin to account for some of the more exotic petroporphyrins, many with molecular weights in excess of 900. Many of these reactions would have been of a radical nature. The environment of a buried sediment (aqueous but excluding oxygen) would lend itself to this kind of chemistry. This would explain how high molecular weight porphyrins could be generated, especially if there is a mechanism of intermolecular transalkylation between porphyrins. Also, oxidations and reductions would have to have gone on side by side. This would mean, for example, that the reduction of vinyl side chains and exocyclic carbonyl groups would probably have occurred in conjunction with the oxidation of the chlorin macrocycle to a porphyrin.

This step highlights an interesting facet of the chlorin/porphyrin interchange. Chlorins are relatively easily oxidised to porphyrins—except for chlorophyll! It takes pretty severe conditions for chlorophyll to undergo oxidation to a porphyrin. Once again, it was R.B. Woodward who provided the most plausible explanation for this phenomenon.[18] One look at the chlorophyll molecule will reveal the crowding of substituents around the γ-*meso*-carbon. The saturation of ring D actually relieves much of the steric hindrance in this region of the molecule by twisting substituents out of the macrocyclic plane. Consequently, aromatisation of ring D in chlorophyll would increase steric hindrance between substituents by making them planar again, even though porphyrins are thermodynamically more stable than chlorins. In other words, it is kinetic factors that maintain the viability of chlorophyll in the face of the increased thermodynamic stability of porphyrins. This means that the aromatisation step in the diagenesis of chloro-

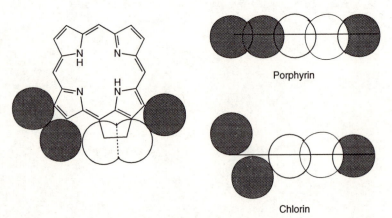

Porphyrin

Chlorin

Fig. 5.11 Steric crowding around C-15 of chlorophyll leading to its stability as a chlorin.

phyll will only occur once some of the bulky substituents have been removed. Lastly, the complexity of the porphyrin fraction increases from sediments to bitumens to kerogens. In other words, the sequence of geo-chemical reactions leading to petroporphyrins converts a highly ordered biomolecule into a collection of geomolecules whose degree of internal molecular order decreases with the amount of thermal stress.

5.6 References

1. See R. Schmid and A.F. Mcdonagh, in *The Porphyrins*, Vol. 6, ed. D. Dolphin, Academic Press, New York (1979), Ch. 5 and references therein; see G.P. Moss, *Pure Appl. Chem.*, (1987), **59**, 779.
2. See A.F. Mcdonagh in ref. 1, Ch. 6.
3. See A. Bennett and H.W. Siegelman in ref. 1, Ch. 7.
4. T. Sjîstrand, *Nature (London)*, (1951), **168**, 729 and 1118; T. Sjîstrand, *Ann. N.Y. Acad. Sci*, (1970), **174**, 5.
5. H. Kayser, *J. Insect. Physiol.*, (1974), **20**, 89.
6. R. Tenhunen, H.S. Marver, N.R. Pimstone, W.R. Trager, D.Y. Cooper, and R. Schmid, *Biochemistry*, (1972), **11**, 1716.
7. I.M. London, R. West, D. Shemin, and D. Rittenberg, *J. Biol. Chem.*, (1950), **184**, 351.
8. F.J. Leeper, *Nat. Prod. Rep.*, (1989), **6**, 171 and references therein.
9. P. O'Carra and E. Colleran, *FEBS Lett.*, (1969), **5**, 295.
10. R. Lemberg, *Biochem. J.*, (1935), **29**, 1322.
11. T. Yoshida and G. Kikuchi, *FEBS Lett.*, (1974), **48**, 256.
12. R. Tenhunen, H.S. Marver, and R. Schmid, *J. Biol. Chem.*, (1969), **244**, 6388; R.K. Kulty and M.D. Maines, *Biochem. J.*, (1987), **246**, 467.
13. R. Tenhunen, M.E. Ross, H.S. Marver, and R. Schmid, *Biochemistry*, (1970), **9**, 298.
14. See S.I. Beale, in *The Biosynthesis of the Tetrapyrrole Pigments*, Ciba Foundation Symposium (April 1993), John Wiley (1993).
15. J.E. Bishop, J.C. Lagorias, J.O. Nagy, R.W. Schoenleber, H. Rapoport, A.V. Klotz, and A.N. Glazer, *J. Biol. Chem.*, (1986), **261**, 6790; L.J. Ong and A.N. Glazer, *J. Biol. Chem.*, (1987), **262**, 6323.
16. P. Nesvadba and A. Gossauer, *J. Am. Chem. Soc.*, (1987), **109**, 6545.
17. B. Kräutler, B. Jaun, K. Bartlik, M. Schellenberg, and P. Matile, *Angew. Chem., Int. Ed.*, (1991), **30**, 1315.
18. See E.W. Baker, and S.E. Palmer, in *The Porphyrins*, Vol. 1, ed. D. Dolphin, Academic Press, New York (1979), Ch. 11.
19. S. Kaur, M.I. Chicarelli, and J.R. Maxwell, *J. Am. Chem. Soc.*, (1986), **108**, 1347; M.I. Chicarelli and J.R. Maxwell, *Tetrahedron Lett.* (1986), **27**, 4653.

6. What happens when it all goes wrong?

6.1 Introduction

We have seen how porphyrins are biosynthesised and, in the last chapter, how they are metabolised. In any normally functioning living system, these processes are usually held in balance, so that a steady-state concentration of any particular substance is obtained. But mistakes do occur and when they do, they lead to a variety of pathological conditions. In this chapter, we shall be examining some of these. We shall also look briefly at what happens when the protein (in particular globin) that surrounds a porphyrin, malfunctions due to genetic misprogramming and does not provide the correct working environment for the heme moiety.

As we saw, the biosynthesis of heme involves a complex series of tightly controlled reactions, starting from glycine. Faults can occur at any stage of the biosynthetic pathway and these generally lead to a build-up of metabolites, which, if not dealt with, sets off one or more of a series of pathological conditions. For example, over-production of metal-free porphyrins, which are deposited in the skin, in the presence of oxygen and light can lead to photosensitivity which in mammals causes disfiguring lesions to exposed areas of the body. Excessive catabolism of porphyrins due to liver malfunction or blockage of the bile duct generates large quantities of bilirubin, which is also deposited in the skin, leading to the orange-yellow colour of jaundice. In newborn babies, for example, where, shortly after birth, the blood undergoes a massive change from fetal to adult hemoglobin, large quantities of red blood cells are broken down leading to over-production of bilirubin. If the baby's system cannot deal with this, jaundice occurs, which, if not treated, can lead to brain damage. The treatment involves exposure to near-UV light which destroys the bilirubin.

Interestingly, both the light sensitivity caused by certain porphyrias, and the use of UV light to destroy bilirubin involve the sensitisation of oxygen into an electronically excited state, where it is much more reactive. Before we examine some of the pathological states that arise from faulty porphyrin metabolism therefore, we shall make a short digression to discuss the photophysics of oxygen.

6.2 The photophysics of oxygen

In Chapter 4, we saw that the ground state electronic structure of oxygen could be described as a triplet (written as $^3O_2$), in which the two highest energy electrons are unpaired, each singly occupying a degenerate antibonding π^*-orbital. We also saw how, in this electronic configuration, oxygen could act as a "sink" for electrons produced via cellular metabolism.

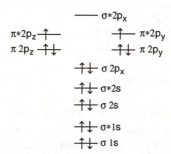

Fig. 6.1 Electronic structure of ground state O_2, $^3O_2$.

Excited states of oxygen are also known. The most accessible of these being one in which the two highest energy electrons are now paired in one of the π^*-orbitals. This is called singlet oxygen and is written as $^1O_2$.[1] Energetically, this state is only 94 kJ mol^{-1} above the ground state so that, in principle, it should be easy to generate singlet oxygen, by exposing triplet ground state oxygen to the appropriate energy radiation (1270 nm is the wavelength of singlet oxygen phosphorescence). Oxygen, however, is transparent at these wavelengths, which means that the energy quanta are simply not absorbed. Photosensitisation[2] will induce the electronic transition from $^3O_2$ to $^1O_2$.

A photosensitiser is a molecule or atom that absorbs radiant energy (for

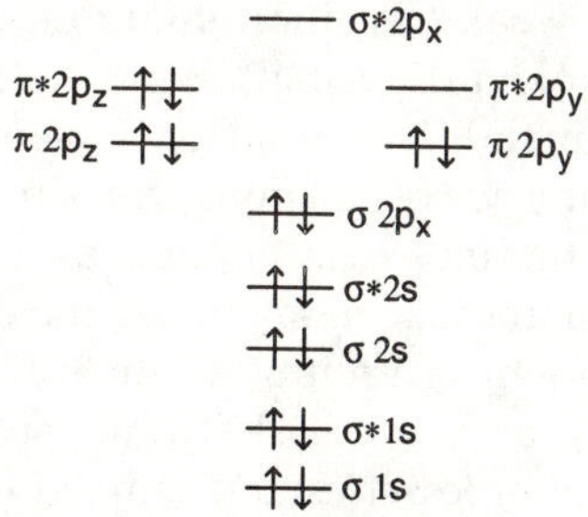

Fig. 6.2 Electronic structure of singlet excited state O_2, $^1O_2$.

our purposes, this is usually visible light), becoming electronically excited in the process, and then, without reacting itself, passes its excitation energy on to another molecule or atom. Chlorophyll in photosynthesis is acting as a photosensitiser, absorbing light energy, becoming electronically excited, and passing this energy on to a sequence of redox reactions that constitutes an electron-transporting chain. After excitation and oxidation, the chlorophyll is reduced by water, returning to its electronic ground state, ready to repeat the operation.

The photosensitisation of oxygen is a slightly different process. The sensitiser molecule, which has a singlet electronic ground state,[1] absorbs light energy and becomes electronically excited. This process occurs very rapidly (about 10^{-15} s) and produces a singlet excited state of the sensitiser, $^{1}S^{*}$, in which the spins of the excited and unexcited electrons are still paired. If the sensitiser is an aromatic molecule, the excited electron will occupy a π^{*}-orbital. The excited electron can lose its energy by falling back down to the electronic ground state, giving up most of the energy it has absorbed in a burst of fluorescence. This is a highly probable process, so that the lifetime of the singlet excited state is usually not more than a few nanoseconds (i.e. of the order of 10^{-9} s). However, under certain circumstances (e.g. if a heavy atom is present) it is possible for either one of the electrons to undergo the "forbidden" process of flipping their spins so that both electrons have parallel spins, giving a triplet excited state of the sensitiser, $^{3}S^{*}$.

In an ideal rigid molecule, such a process would indeed be forbidden, i.e. just could not happen. The selection rules that govern electronic transitions make this clear.[3] These rules arise from the mathematical treatment (based on solutions to the Schrödinger equation) of the behaviour of electrons in molecules. The success of this treatment depends on its power to predict the intensities of bands in the spectra of molecules (e.g. UV–visible for electronic transitions and infrared for molecular vibrations) and to explain why certain transitions do not appear or are weak. The mathematical technique that is used is based on matrix algebra and requires the determination of a matrix element called the dipole moment operator. If this turns out to be zero for a particular transition, then that transition does not occur, i.e. is forbidden. The fact that a weak transition might occur, i.e. the selection rule for that transition is broken, is usually because the molecular wavefunctions that were used in the derivation of a particular selection rule are approximate. Thus, many wavefunctions, for simplicity's sake, are calculated on the basis that the molecule vibrates as a harmonic oscillator and that the spin and orbital angular momenta of the electrons in the molecule are completely independent of each other. In reality, molecules behave as anharmonic oscillators, and their electronic spin and orbital angular momenta are, to a greater or lesser extent, coupled (spin–orbit coupling); this becomes more important the heavier the molecule. Consequently, selection

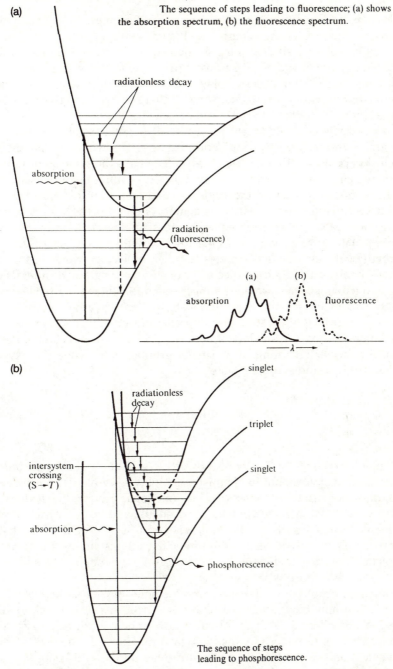

(a)

The sequence of steps leading to fluorescence; (a) shows the absorption spectrum, (b) the fluorescence spectrum.

radiationless decay

absorption

radiation (fluorescence)

(a)

(b)

absorption

fluorescence

$\longrightarrow \lambda \longrightarrow$

(b)

singlet

radiationless decay

triplet

singlet

intersystem crossing $(S \rightarrow T)$

absorption

phosphorescence

The sequence of steps leading to phosphorescence.

Fig. 6.3 Excitation of a molecule leading to (a) fluorescence and (b) phosphorescence.

rules are broken and forbidden behaviour becomes, to a small extent, allowed. The greater the extent of spin–orbit coupling, the more chance of excited triplet state species being formed.[4] The situation is rather similar to fouls in a game of soccer. The rules of the game forbid such actions so that not much of the 90 minutes of play is taken up with it. However, fouls do occur (the rules are broken), and some of that activity even goes undetected (or is ignored)! But back to the world of atoms and molecules.

When a triplet excited state is formed, it is not possible for the excited electron to return directly to its ground state as this would violate the Pauli exclusion principle (that no two electrons can have the same quantum numbers, which is another way of saying that they cannot be in the same place at the same time). Before the triplet excited state can lose its energy, the electrons must first pair their spins again. As this process has a low probability, it means that, compared with the singlet excited state, the triplet excited state can survive for a long time (anything from microseconds to seconds), unless, that is, it meets another molecule which has a lower energy triplet state. Such a molecule is, of course, oxygen. The interaction of the two triplet species provides a mechanism for the excited triplet state to fall back down to its singlet ground state without violating any fundamental cosmic principles. The triplet excited state of the sensitiser passes its excitation energy (not an electron) on to the triplet ground state of the oxygen, the former returning to its singlet ground state, while the oxygen is converted to a singlet excited state.

$$^1S + h = {}^1S*$$

$$^1S* = {}^3S*$$

$$^3S* + {}^3O_2 = {}^1S + {}^1O_2$$

Singlet oxygen is a far more powerful oxidising agent than ordinary ground state oxygen, but its chemistry is highly selective. Thus, singlet oxygen does not react with most organic functional groups, but it does react readily with conjugated polymeric systems and polynuclear aromatic systems. In many of its reactions, it behaves as a dienophile.[5] We shall see how this reactivity can be a boon, protecting infants against the ravages of neonatal jaundice, and how it can cause severe problems to sufferers of certain kinds of porphyria.

But first, a note of caution. Although singlet oxygen is highly reactive, doubts are now beginning to be expressed over just how much of singlet oxygen chemistry is due to $^1O_2$ and how much to singlet oxygen precursors.[6] Also, in water (i.e. the biological medium) the lifetime of singlet oxygen is only 2 s so that reactions attributed to it would be localised near to the site of its generation.

6.3 Neonatal hyperbilirubinaemia—newborn jaundice

For most human adults, the four globin chains of the hemoglobin tetramer are two α globins and two β globins, each enfolding a heme moiety.[7] This regular form of hemoglobin is called hemoglobin A. But not everyone's hemoglobin is the regular variety. There are a class of diseases called thalassaemias, in which imbalances occur in the ability to produce the various globin chains. Also, we do not arrive in this world with fully formed, adult hemoglobin. The blood of the human fetus, for example, contains so-called fetal hemoglobin, known as hemoglobin F, which has a greater oxygen-carrying power than hemoglobin. A. Hemoglobins A and F have similar structures, in that they consist of globin tetramers, but in F the β chains are replaced by γ chains which, although having the same number of amino acids (146), possess a slightly different amino acid sequence.[8]

Up to 50% of a newborn infant's blood can consist of the F-type hemoglobin, but it is normally replaced with hemoglobin A, soon after birth. In certain congenital anaemias hemoglobin F fails to disappear, but for the majority of newborns, the replacement of hemoglobin F with A involves a massive destruction of fetal heme, and its conversion to large quantities of bilirubin (hemolysis). This can overload the body's mechanisms for excreting bilirubin, and jaundice (i.e. perceptible yellowing of the skin) occurs due to the presence of mainly free bilirubin and some bilirubin glucuronides in the blood serum and plasma.

There are of course, many other causes for jaundice in newborns (notably due to incompatabilities between the blood types of the mother and fetus. Thus, if the mother is rhesus negative, she will make antibodies against her own baby if the baby's blood, and that of the father, is rhesus positive, and mixes with hers during pregnancy—if that happens, the mother's antibodies will begin to destroy the baby's red blood cells and the baby will be born jaundiced.) What they all have in common is a build-up of bilirubin in the system, either because of blockages of the bile duct, or more importantly because excessive hemolysis is occurring. For newborn infants, this is particularly dangerous because the blood–brain barrier has not yet had time to develop. The bilirubin, therefore, can accumulate in a part of the brain called the basal ganglia. This produces a neurological condition called kernicterus, a type of brain damage characterised by muscle spasm leading to arching of the back and clenching of the fists. Intellect and hearing are generally also impaired. Fortunately, it is possible to counter the effects of excessive bilirubin build-up in newborns by adopting a very simple procedure. The child is simply exposed to near-UV light. This modern version of Spartan exposure of the newborn, has the magical effect of destroying the bilirubin via its interaction with light and oxygen.[9]

Bilirubin absorbs light energy and is excited to a singlet excited state. This is followed by transformation to the triplet state. In the presence of oxygen, there follows the series of changes described in the previous section, which eventually leads to the generation of singlet oxygen. Because this is generated in the vicinity of bilirubin (which also happens to be the nearest substance whose molecules contain reactive double bonds), in the presence of near-UV light and oxygen, bilirubin ends up photosensitising its own destruction.

6.4 Faulty hemoglobins

6.4.1 Methemoglobinaemia

We have seen how in hemoglobin, the heme units are held within hydrophobic pockets of amino acids inside the globin protein chains. Furthermore, each heme iron is coordinated, in its fifth and sixth positions, to two imidazole side chains of histidine amino acid residues. One of these imidazole units is further from the heme iron than the other, so allowing space for oxygen to bind to the iron. This has the effect of converting the Fe(II) cation into its low-spin electronic configuration. In this supramolecular set-up (i.e. coordination to a porphyrin, two imidazoles, and in a hydrophobic environment), the iron is less capable of being oxidised to Fe(III). If this occurs, however, methemoglobin is produced which is useless for transporting oxygen.[10]

There are several reasons why this might happen. In fact it happens every day to all of us, about 1% of the total circulating hemoglobin being spontaneously converted to methemoglobin. In normal adults, there are enzymatic protection mechanisms that ensure the oxidised heme (hemin) is reduced back again. However, certain drugs, aniline dyes, and food preservatives (e.g. nitrates and nitrites) are capable of converting heme to hemin via the bound oxygen. Methemoglobinaemia results and can lead to death if severe enough. But there can be more deep-seated reasons for the onset of this condition.

Clearly, the amino acid sequence of the enfolding globin chains has a crucial effect on the chemistry of the iron. This sequence is determined genetically and many variations of some of the amino acids have been recorded; we have just been introduced to one in fetal hemoglobin. Several hemoglobins are known where genetic variation takes place at one of the histidines whose imidazole side chains coordinate to iron. The histidine is replaced by a tyrosine residue which has a phenol side chain. Alternatively, a valine residue near the iron binding site is converted to a glutamic acid residue.[11]

In the first case, the oxygen of the phenol side chain coordinates to the

Abnormal hemoglobin	Position and normal residue	Replacement
	α chain	
I	16 Lys	Glu
G$_{Honolulu}$	30 Glu	Gln
Norfolk	57 Gly	Asp
M$_{Boston}$	58 His	Tyr
G$_{Philadelphia}$	68 Asn	Lys
O$_{Indonesia}$	116 Glu	Lys
	β chain	
C	6 Glu	Lys
S	6 Glu	Val
G$_{San Jose}$	7 Glu	Gly
E	26 Glu	Lys
M$_{Saskatoon}$	63 His	Tyr
Zürich	63 His	Arg
M$_{Milwaukee}$	67 Val	Glu
D$_{Punjab}$	121 Glu	GLN

Fig. 6.4 Changes in amino acid sequences of some abnormal hemoglobins.

iron. This has the effect of decreasing the strength of the ligand field around the iron so that the low-spin configuration is less accessible. The iron is then easier to oxidise to Fe(III) by the coordinated oxygen, and less amenable to reduction back to Fe(II) by the cells' enzyme-based reductive mechanisms. The iron is locked into the Fe(III) oxidation state, with concomitant inability to reversibly bind oxygen. In the second case, the nearby glutamic acid residue has two effects. Its carboxylic oxygen competes with the histidine imidazole side chain for coordination to the iron, and it makes the pocket containing the heme less hydrophobic. Any water that finds its way into the pocket will also compete with imidazole for heme iron coordination. Both effects result in easier oxidation of Fe(II) to Fe(III) and locking of the iron into the Fe(III) oxidation state, as before.

Fortunately, in such cases of genetically induced deviations, not all of the person's hemoglobin is abnormal. Only one or two of the globin chains in a particular hemoglobin molecule are affected, and the abnormal hemoglobin (called hemoglobin M) usually makes up only a proportion of the total hemoglobin content. Consequently, a relatively normal life can be led.

However, the presence of methemoglobin may sometimes lead to skin pigmentation, a condition known as cyanosis. This is because methemoglobin gives blood a brown colour.

Children with hemoglobin M show a difference in the time of onset of cyanosis depending on whether the genetic abnormality is expressed in the α- or β-globin chain. If in the former, cyanosis is observed from the first day of life, because the α-globin chain is common to fetal and mature hemoglobins. However, if the abnormality is expressed in the β chain, then cyanosis will not become noticeable until enough of the fetal hemoglobin (which contains γ-globin chains in place of β chains) has been replaced with adult hemoglobin containing the abnormal β chains.

6.4.2 Sickle-cell hemoglobin

This is the most important of the abnormal variants of hemoglobin and the sickle-cell gene is found mainly in people of African origin. What happens is that the usually circular red blood cells have a tendency to collapse into a sickle shape when the oxygen concentration in the blood is low. To understand the reason for this, it is first necessary to realise that in a normal red blood cell, the concentration of hemoglobin is so high that the intracellular fluid is highly viscous and near to crystallisation.[12]

At high oxygen concentrations, oxyhemoglobins of both the normal and the sickle-cell varieties (hemoglobin S), have the same solubility in the intracellular fluid of the red blood cell. But, when the oxyhemoglobin deoxygenates, it has been found that the hemoglobin S is much less soluble than normal hemoglobin, and crystallises out of the intracellular fluid, forming one-dimensional crystals called *tactoids*. This causes the red blood cell to collapse and to sickle. The crystallisation of the deoxyhemoglobin S makes it harder to pick up oxygen than with normal deoxyhemoglobin, so that the higher the concentration of hemoglobin S, the more difficult it is for oxygen to be efficiently transported. Collapsed, sickled red cells can regain their shape on oxygenation, but continued sickling on deoxygenation eventually causes irreversible sickling of the cell due to membrane damage.[13]

The sickling of red blood cells slows down the blood circulation because these cells block up the smaller veins. This is one of the causes of the anaemia. The slower the blood circulation round the body, the less chance the sickle cells have of oxygenating in the lungs, so that a vicious circle is set up. Less oxygen means more sickling, which slows the blood, leading to less oxygen and more sickling, and so on. The sickled cells have a short life span (about two days), which is another cause of the anaemia.

The crystallisation of hemoglobin S at low oxygen concentrations is caused by a genetic variation of one or more of the amino acids in the β-globin chains. The classic sickling mutation is replacement of the hydro-

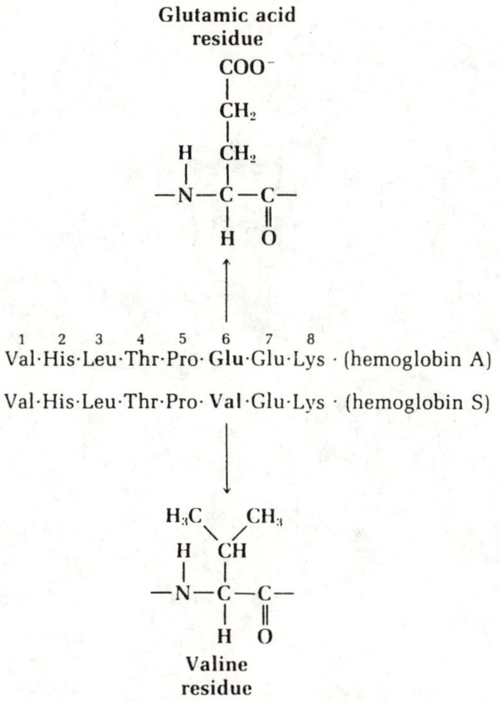

Glutamic acid
residue

$$COO^-$$
$$|$$
$$CH_2$$
$$|$$
$$H \quad CH_2$$
$$|\quad\ |$$
$$-N-C-C-$$
$$|\quad\ \parallel$$
$$H \quad O$$

1   2   3   4   5   6   7   8
Val·His·Leu·Thr·Pro· Glu·Glu·Lys · (hemoglobin A)

Val·His·Leu·Thr·Pro· Val ·Glu·Lys · (hemoglobin S)

$$H_3C \quad CH_3$$
$$\diagdown \diagup$$
$$H \quad CH$$
$$|\quad\ |$$
$$-N-C-C-$$
$$|\quad\ \parallel$$
$$H \quad O$$

Valine
residue

(a)

Fig. 6.5 The genetic defect in sickle-cell hemoglobin. As the result of a mutation in the β chain gene, the glutamic acid residue normally present in the 6-position of the β chain of hemoglobin A is replaced by a valine residue. This replacement results in the loss of one negative charge in each of two β chains.

philic glutamic acid residue at position 6 in the β chains by a hydrophobic valine residue.[13] This makes deoxyhemoglobin S slightly less soluble in the intracellular fluid, but also allows its molecules to fit together more easily. The result is the formation of right-handed helical stacks of deoxy-hemoglobin S molecules and the formation of tubular crystals.

The sickling of red cells becomes a life-threatening anaemia when the majority of the hemoglobin is of the S variety. Any reduction in oxygen tension can bring on a sickling crisis, which can lead to death. Anaesthetics have to be avoided for this reason. Likewise, high altitudes in unpressurised aircraft. The sickle-cell trait occurs when only 25–45% of the hemoglobin is S. People with the sickle-cell trait generally lead normal lives.

Does the sickling gene serve any purpose, or is it just a tragic genetic mistake? One thing the gene does do is to give protection against some of the more virulent forms of malaria.

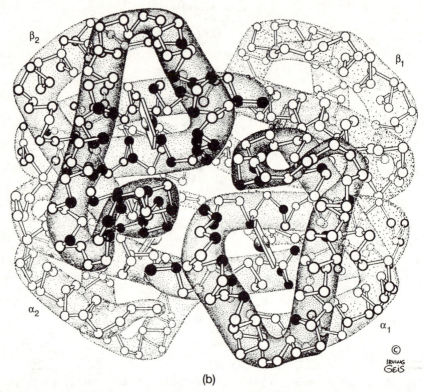

(b)

Fig. 6.6 Positions of 163 mutations (black circles) in human hemoglobin observed as of 1979. There are 105 mutations in the β chains and 58 in the α chains. Mutations occurring near the heme groups are most likely to lead to serious defects in hemoglobin function.

6.5 The porphyrias

The things that can go wrong that we have dealt with up to now involve either the protein around the heme (leading to heme malfunction as an oxygen carrier in hemoglobin), or the consequences of excessive heme catabolism. There is another class of diseases, usually (but by no means always) genetically inherited, which are associated with a partial defect in one of the many enzymatic steps of heme biosynthesis. This family of diseases goes under the generic name of *porphyrias*, and their main clinical manifestations are an increase in the excretion of porphyrins and other metabolites in the urine and feces. The symptoms that are usually presented involve a sensitivity to light in exposed areas of skin, and, in extreme cases, to photocutaneous lesions and neurological dysfunction.[14]

In mammalian tissues, usually the liver and the bone marrow, heme is synthesised in eight discrete enzyme-catalysed steps. These steps begin within the mitochondrion of the cell with the coupling of succinyl CoA with glycine to give 5-aminolaevulinic acid or ALA. The synthesis of ALA is under tight control by two mechanisms. First, the enzyme responsible for ALA synthesis, ALA synthetase, is itself produced outside the mitochondrion in the aqueous part of the cellular cytoplasm (called the cytosol). Its transport across the double membrane between the cytoplasm and the mitochondrial matrix is closely regulated. Secondly, the final product, heme, acts to inhibit ALA synthetase. However, a defect in ALA synthetase is not usually considered a porphyria because the substrates that accumulate (succinate and glycine) are intermediates for other biosynthetic pathways. Heme biosynthesis then moves outside the mitochondrion into the cytosol, with the ALA dehydrase-catalysed condensation of two ALAs to give porphobilinogen, PBG. A deficiency in this enzyme leads to a porphyria called, naturally enough, *ALA dehydrase deficiency*. This is characterised by ALA appearing in the patient's urine, accompanied by neurological dysfunction, of which the main symptom is abdominal pain. However, in acute conditions, psychiatric disorders can occur.

Psychiatric disorders (in so far as they can be explained by imbalances in neurochemicals) that accompany some of the porphyrias may be caused by a build-up in levels of ALA, which bears a structural resemblance to the neurotransmitter GABA (γ-aminobutyric acid), and so could act as a neurotoxin. The heme deficiency that is brought on by the porphyrias, can lead to a reduction in the activity of hepatic (liver) enzymes that require heme. For example, reduction in the level of hepatic tryptophan pyrrolase activity leads to a build-up in levels of the amino acids tryptophan and 5-hydroxytryptophan. Thus, a heme-deficient state in the liver could produce biochemical abnormalities capable of leading to neurological dysfunction, while heme deficiency in nerve tissue could directly alter nerve function. This has led to the treatment of severe neurological dysfunction by intravenous administration of heme compounds.

The next stage in heme biosynthesis involves the coupling of four PBG units head-to-tail to a linear tetrapyrrole, via the enzyme PBG deaminase. The linear tetrapyrrole is then cyclised regiospecifically to uroporphyrinogen III via the enzyme uro'gen III cosynthase. Without this enzyme, the linear tetrapyrrole spontaneously cyclises to uroporphyrinogen I. *Acute intermittent porphyria* results from problems with PBG deaminase activity. This entails neurological dysfunction and the presence of ALA and PBG in the urine. However, a lack of uro'gen III cosynthase leads to a condition called *congenital erythropoietic porphyria*, characterised by a build-up of uroporphyrinogen I, which is not biosynthesised further. This is excreted in the urine and feces and also appears in blood serum. Patients with this disorder are light sensitive.

From uro'gen III, heme biosynthesis, which at this stage is still located in the cytosol, proceeds to coproporphyrinogen III by decarboxylation of the four carboxymethyl substituents to methyl groups via the enzyme uro'gen III decarboxylase. Defects in the production and function of this enzyme lead to two conditions called *porphyria cutanea tarda* (PCT) and *hepatoerythro-poietic porphyria* (HEP). In these conditions, the particular gene that pro-grammes for the production of uro'gen III decarboxylase, inserts a valine (PCT) or glutamate (HEP) amino acid residue where there ought to be a glycine. The result is an enzyme protein with a severely shortened half-life; in the case of PCT 4 hours instead of 104 hours. Consequently, there is a concomitant decrease in the amount of this enzyme in the cytosol, with a build-up of porphyrinic metabilites. The siting of these metabolic build-ups varies between PCT and HEP. In the former, porphyrin over-production occurs in the liver with uroporphyrin and coproporphyrin appearing in the urine and feces, respectively. In the latter, porphyrin over-production occurs in the bone marrow as well as the liver, with uroporphyrin and coproporphyrin appearing in the urine and feces, as before, but some zinc protoporphyrin appearing in red blood cells. In both PCT and HEP, por-phyrins are also deposited in the skin because the major clinical finding of both of these conditions is photocutaneous lesions.

The next enzymatic step that can be affected by genetic abnormality is the oxidative decarboxylation of copro'gen III to proto'gen IX via the enzyme *copro'gen oxidase*. It is at this stage that heme biosynthesis begins to move back inside the mitochondrion, because the enzyme is located on the outer surface of the inner mitochondrial membrane. Defects in this enzyme lead to a condition called *hereditary coproporphyria*, which produces both photo-cutaneous lesions and neurological dysfunction. It is characterised by the appearance of ALA, PBG, and coproporphyrin in the urine, with the latter appearing in the feces as well. Proto'gen IX is then enzymatically oxidised to protoporphyrin IX via a proto'gen oxidase, which is an integral protein of the inner mitochondrial membrane. Genetically produced defects in this enzyme lead to a condition called *variegate porphyria*, exhibiting both the usual symptoms, of photocutaneous lesions and neurological dysfunction. Once again ALA, PBG, and coproporphyrin are found in the urine, but this time protoporphyrin is found in the feces as well as the bile.

Iron insertion is the final step of heme biosynthesis, and this is controlled by the enzyme *ferrochelatase*. This enzyme is located on the inner face of the mitochondrial inner wall and its failure to function properly leads to a con-dition called *erythropoietic protoporphyria*, which, because protoporphyrin IX builds up in the skin, leads to light sensitivity. Protoporphyrin is also found in the red blood cells, bile, and feces. The three porphyrias that arise due to genetic malfunction of the last three enzymes in heme biosynthesis have not yet been tracked down to specific genetic abnormalities.

The photosensitivity caused by some of the porphyrias is due to the photophysical properties of the porphyrins circulating in the red blood cells or actually deposited in the skin. Metal-free porphyrins, or porphyrins containing metal ions that do not interact with the porphyrin π-system, are photoexcited by near-ultraviolet light into the singlet excited state, which, as we saw previously, can undergo intersystem crossing to the triplet excited state. It is in this excited state that the porphyrin can bring about the photosensitisation of oxygen to the singlet excited state. The singlet oxygen will then attack sensitive molecules with double bonds, e.g. those in the cell membranes. The cells are wrecked, leading to the photocutaneous lesions mentioned earlier.

Other conditions, besides the porphyrias, are known that lead to small increases in urinary porphyrin excretion. Clinically, these are grouped under the heading of secondary porphyrinuria and they usually arise from certain anaemias and liver/bile disorders. Misdiagnosis with porphyria is avoided by measuring the amount of ALA and PBG in the patient's urine, as the levels in secondary porphyrinuria are normal.

Lead poisoning can affect the heme metabolic pathway at several sites, and this generally leads to an increase in the zinc protoporphyrin level of red blood cells. In fact, by measuring the fluorescence emission from red cell zinc protoporphyrin, it is possible to screen large populations for lead poisoning.[15] Other substances that can affect the liver, producing porphyria-like symptoms, are barbiturates, DDT, and alcohol.

The porphyrias are, thankfully, rare. But it is just possible that observation of the manifestations of these conditions in earlier times may have given rise to mythologies such as the vampire and the werewolf. Let us examine how, by painting a picture of a sufferer with an acute porphyric condition.

1. Such a person would be extremely photosensitive, probably with horribly disfigured hands and face (i.e. the parts of the body exposed to daylight).

2. As a result, the person may choose to go out only at night, and possibly, being mentally disturbed, exhibiting abnormal behaviour. In a superstitious community they would most likely be ostracised and persecuted for being possessed by "evil spirits".

3. In certain porphyrias, not only would the urine be stained red, from the passage of porphyrins (one of the supposed hallmarks of a werewolf[16]), but porphyrins are deposited in the teeth, leading to their being stained red. In some cases, gum tightening and recession would give the impression of elongated canines.

4. Hirsutism, the growing of excessive body hair, also occurs in certain porphyrias and the occurrence of jaundice due to acute hepatic dysfunction, would give the sufferer a distinctive sallow complexion.

We can start to see parallels between the description of a supposed vampire or werewolf, and that of an acute porphyria sufferer.

What about the vampire's aversion to garlic? This bulb is well known for its alleged blood-purifying character. Garlic could potentiate the heme oxygenase enzyme with its cytochrome P450-like activity, which is used to break down old red blood cells by splitting open the heme molecule to give bile pigments and releasing the iron atom. Anyone with a severe disorder of heme metabolism would have a vested interest, therefore, in not ingesting a substance that would accelerate the breakdown of any heme they might posses. It is also not inconceivable that an acute porphyria sufferer might have a taste for blood. This would be a ready source of heme, which would survive the protein-digesting trip through the stomach and intestines. In just the same way that some of the well-known cravings of pregnant women can be linked to direct bodily requirements for certain minerals, etc., it is possible that the demand for heme could lead to ingestion of fresh blood.

6.6 References

1. H.H. Wasserman and R.W. Murray (eds), *Singlet Oxygen*, Academic Press, New York (1979) and references therein.
2. G. Bock and S. Harnett (eds), *Photosensitising Compounds; Their Chemistry, Biology, and Chemical Uses*, Ciba Foundation Symposium, No. 146, Wiley, Chichester (1989).
3. C.N. Banwell, *Fundamentals of Molecular Spectroscopy*, 3rd edn, McGraw-Hill, London (1983).
4. W.G. Richards and P.R. Scott, *Structure and Spectra of Molecules*, Wiley, Chichester (1985), p. 103.
5. A.A. Frimer (ed.), *Singlet Oxygen*, Vols 1–4, CRC Press, Boca Raton, FL (1985).
6. D.T. Sawyer, *Oxygen Chemistry*, Oxford University Press, Oxford (1991), p. 157.
7. See J. Baggott, in *Textbook of Biochemistry*, 3rd edn, ed. T. Devlin, Wiley Liss, (1992), p. 1030.
8. H. Lehmann and R.G. Huntsman, *Man's Hemoglobins*, North-Holland, Amsterdam (1974).
9. J. Dalton, L.R. Milgrom, and R. Bonnett, *Chem. Phys. Letts.*, (1979), **61**, 242 and references therein.
10. See C.E. Castro, in *The Porphyrins*, Vol. 5, ed. D. Dolphin, Academic Press, New York (1978), p. 18.
11. See ref. 8, p. 207.
12. See ref. 8, p. 144.
13. A.L. Lehninger, *Principles of Biochemistry*, Worth, New York (1982), p. 194; ref. 8.
14. See L. Eales in ref. 10, Vol. 6.
15. A.A. Lamola, M. Joselow, and T. Yamane, *Clin. Chem.*, (1975), **21**, 93.
16. L. Illis, *Proc. Roy. Soc. Med.*, (1964), **57**, 23.

7. Porphyrins for the future

7.1 The phthalocyanines

7.1.1 Introduction

The porphyrins we have dealt with so far are naturally occurring, and are involved in the light-harvesting reactions of photosynthesis and the electron- or oxygen-transporting and storage functions of cellular metabolism. Synthetic porphyrins are also useful, and, as our knowledge progresses, the number of applications is expanding. Sophisticated porphyrin derivatives are already being tried out in laboratories around the world in a wide variety of applications, including medicine, electronics, and alternative energy generation. However, before we examine some of these new applications, we should start with a lucky fluke that led to one of the most successful families of synthetic pigments ever invented—the phthalocyanines.[1] These blue to green pigments now have a wide variety of uses, including artists' colours, printing inks (especially for bank notes), surface coatings, paper dyes, floor coverings, and even as high temperature lubricants for space vehicles. But the success of synthetic pigments like the phthalocyanines is relatively recent.

Before the chemical industry really got into stride, during the second half of the nineteenth century, dyes were extracted from natural sources. We saw in the first chapter how the ancients extracted purple dyes from sea molluscs. Plants, too, were a valuable source of colours, in particular, the dark blue dye indigo, which was extracted from the indigo plant. In the leaves of the indigo plant, there exists the indole derivative called indican, which is the glucoside of 3-hydroxyindole (indoxyl). Hydrolysis of indican, followed by air oxidation, led to cleavage of the glucose residue (to release indoxyl) and to oxidative dimerisation to the dark blue indigo. This process was carried out in the presence of the cloth to be dyed in large vats—hence the family name for dyes applied in this way; vat dyes.

Before 1897, indigo was obtained from plants grown in Java and Bengal. In that year the then young chemical company BASF, based at Ludwigshafen in Germany, started to manufacture synthetic indigo. The starting material for their synthesis was naphthalene, just one of the many products obtained from coal tar. There are now simpler ways of doing it, but BASF oxidised naphthalene using a mixture of sulphur trioxide and mercuric sulphate to give orthophthalic acid. This was then reacted with

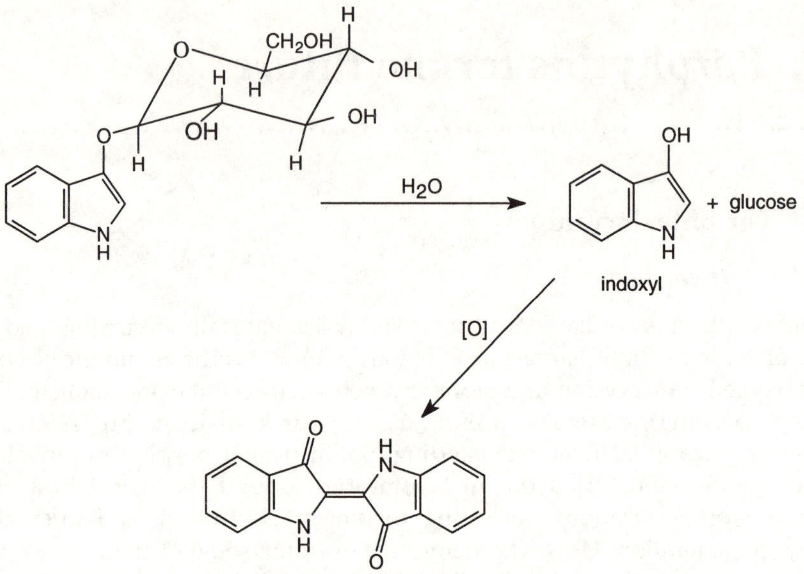

Fig. 7.1 Production of indigo from indican.

ammonia to give phthalimide, a key intermediate in this and several other synthetic processes (including the production of amines and antibiotics). Further reaction with sodium hypochlorite, followed by chloroacetic acid, alkali fusion, and aerial oxidation led to synthetic indigo, a synthetic and commercial triumph for BASF.

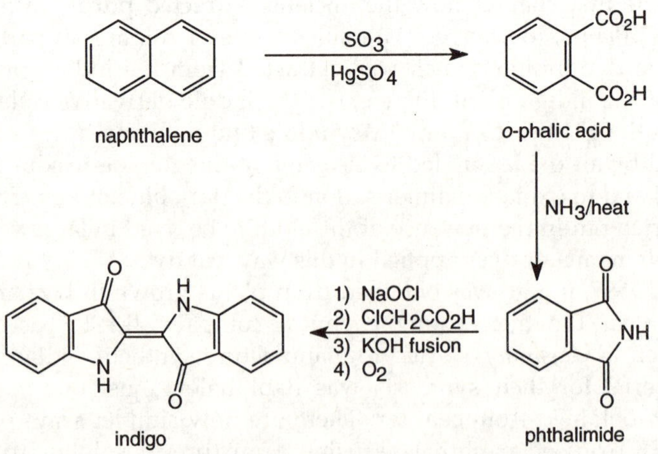

Fig. 7.2 BASF route to indigo from naphthalene.

7.1.2 *The dye is cast*

Thirty years later and a geographical shift to Scotland, where the Scottish Dye Works (later to become part of the multinational ICI), in 1928, were routinely preparing phthalimide from phthalic anhydride (like phthalic acid but minus a molecule of water) and ammonia. They started to use iron pots for the synthesis and found that their phthalimide was inconveniently contaminated with a deep blue highly stable and insoluble pigment. A year earlier, in 1927, a German chemist called De Diesbach obtained a similar blue compound by reacting phthalodinitrile with a copper salt. It took the skill of the British chemist Sir Patrick Linstead, then at London's Imperial College, in 1932, to show that both the Scottish Dye Company's irritating blue pigment and De Diesbach's blue compound were the iron and copper complexes, respectively, of phthalocyanine.

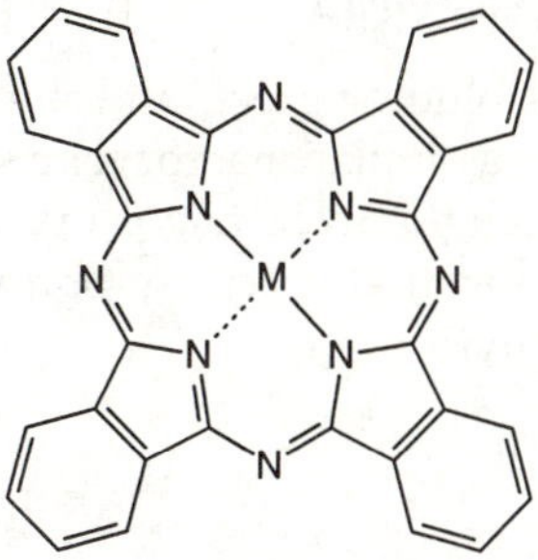

M = Fe(III) or Cu(II)
X = Cl

Fig. 7.3 Iron and copper phthalocyanines.

Just how stable phthalocyanines are can be gauged from the fact that the iron cannot be removed even with concentrated sulphuric acid, and that the blue pigment is stable right up to 500 °C. The potential of phthalocyanines was immediately recognised and a patent was filed describing what Linstead later recognised as iron(II) phthalocyanine, or *iron(II) tetrabenztetraazaporphyrin*, to give it its correct chemical name. This led to a flurry of activity to find ways of efficiently preparing metal phthalocyanines and the best Linstead preparation of copper phthalocyanine—from the fusion of phthalodinitrile with copper bronze at 190–270 °C—gave the product in 75–90% yield. These chemists also realised that earlier reports of insoluble blue-green materials from heating compounds similar to phthalodinitrile (*o*-cyanobenzamide, for example), some stretching back to 1907, must have been phthalocyanines that went unrecognised.[2] ICI patented copper phthalocyanine under the name of Monastral Blue 5025 and the cheapest way of making it industrially is to heat phthalic anhydride with urea in the presence of copper(I) chloride and a catalyst, such as aluminium oxide.

Introducing chlorine atoms into the fused benzene rings around the phthalocyanine macrocycle leads to a change in the colour of these pigments from blue to green. Other groups were introduced, for example sulphonate groups and quaternary ammonium groups, which make the phthalocyanines soluble in water. So the insoluble pigment that could be dispersed in a liquid to give a paint, was converted into a dye. In fact, by using the quaternary ammonium salts as water-solubilising groups, a material or paper could be dyed, and then by subjecting the material to a steaming process, the groups are split off, leaving the pigment perfectly and irreversibly dispersed in the fibre of the material. This was the basis of ICI's successful dye, Alcian Blue.

7.1.3 Why phthalocyanines are blue

Why are phthalocyanines blue or green and not red like the porphyrins? The phthalocyanines have a similar macrocyclic structure to porphyrins so you might expect the same light-absorbing properties. They are not and one look at the UV–visible absorption spectrum of a phthalocyanine will show how it differs from a porphyrin.

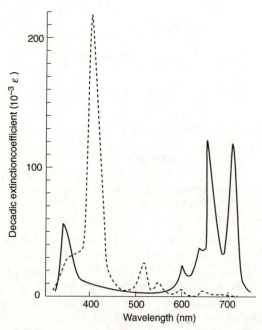

Fig. 7.4 Typical porphyrin and phthalocyanine UV–visible spectra.

Like the porphyrins, phthalocyanines absorb in the near-ultraviolet and visible region, but the intensities of the absorptions are entirely different. It is the visible absorption bands that are more intense than the near-ultraviolet bands, not the other way round, as with porphyrins. The reasons for this are perturbations to the phthalocyanine π-system caused by, (a) the nitrogen atoms in the *meso*-positions (they are more electronegative than carbon atoms so that they tend to attract π-electron density towards themselves) and, (b) the fused benzene rings on the pyrrole β-positions, which extend the π-system (they increase the size of the electron "box").

In Chapter 3, we saw how the four-orbital model of Martin Gouterman (which explains the main facts of porphyrin UV–visible spectra by considering only electronic transitions between the HOMOs and LUMOs) predicted an intense near UV band and weak visible bands for porphyrin spectra. The four-orbital model also predicts that perturbations to the energies of the HOMOs (which have a_{1u} and a_{2u} symmetry) and LUMOs (which both have e_g symmetry), caused by replacing the porphyrin *meso*-carbons with nitrogen atoms and fusing benzene rings onto the pyrrole β-positions vastly increases the probability of forbidden transitions between these orbitals.[3] The net effect is to increase the intensity of the visible bands at the expense of the near-UV band. These visible bands are between 600 and 700 nm, which is the red end of the visible spectrum. Red light, therefore, is absorbed, while the blue and green parts of the visible spectrum are less well absorbed. Consequently, the phthalocyanines appear blue to green, depending on the exact positioning of these visible absorption bands.

Apart from their use as dyes and pigments, phthalocyanines are also being used as catalysts for oxidations, high temperature lubricants, semiconductors, and liquid crystals. Their robust chemical structures, which make them, as dyes and pigments, "fast" to light and heat stable up to 500 °C, means that phthalocyanines will find many more applications in the future. In a later part of this chapter, we shall see how phthalocyanines are being used as novel electrical conductors.

7.2 Porphyrins and alternative energy

7.2.1 Introduction

Humans have always used fire, if only to keep warm. With the industrial revolution, combustion of coal provided the energy necessary to make steel. In this century, oil and gas have begun to replace coal as the prime combustible material, especially for electricity generation. However, the problem of how to manage the products of combustion is now one of the most important environmental issues. Increasing carbon dioxide concentrations in the atmosphere from the burning of fossil fuels, are thought to be one of the main causes of global warming, while emissions of sulphur dioxide and oxides of nitrogen contribute to acid rain which wipes out forests and kills aquatic life in lakes and rivers.

Nuclear "burning"—the atomic fission of uranium and plutonium, which releases much larger quantities of energy than ordinary chemical combustion—has yet to fulfil its promise, not only in terms of cost but also in trust. How to dispose of dangerous radioactive waste safely and cleanly is still an insoluble problem.

7.2.2 The hydrogen economy

Hydrogen could be the ideal answer to these problems. When it is burnt (delivering more energy per mole than hydrocarbon fuels or coal), there is no waste, as the product of combustion is water. And if water could be used as a source of hydrogen, then we should possess a truly clean, renewable energy source in virtually inexhaustible supply.

The practical realisation of such a dream is inevitably beset with problems. First, hydrogen has a bad image. This may seem surprising in the light of global warming and acid rain, but the vision of burning airships,

Fig. 7.5 Newsreel picture of The Hindenburg burning in New York, 6 May 1937. [Camera Press, London.]

Fig. 7.6 A hydrogen car. [Courtesy of BMW.]

like the Hindenburg (which to keep aloft, required millions of cubic feet of flammable hydrogen) which caught fire at its New York mooring in 1937, is still a potent symbol of the dangers of this gas. That hydrogen burns with a "safer" flame than hydrocarbons is ignored, as is the fact that many of the passengers in the stricken airship had time to jump to safety before the main diesel fuel that powered the ship's engines, caught fire. After that, there were no more survivors.

Secondly, hydrogen is not as easy as other gases to transport. Liquid hydrogen has to be stored in large vacuum flasks, and hydrogen gas can be combined with certain compounds to form hydrides, which are solid and give up their hydrogen easily. Piping it presents difficulties because hydrogen molecules are so small, they can easily diffuse through the material of the pipe. Nevertheless, vehicles running on hydrogen gas, which they burn in place of hydrocarbons, have been on the roads. BMW, for example, have several experimental cars that run on hydrogen stored in vacuum flasks in their car boots.

The main problem, however, is to produce hydrogen cheaply enough to compete with fossil fuels. The most promising way is to "crack" water, but that usually requires electricity to decompose the water into hydrogen and oxygen. Only if there is a cheap source of electricity, such as hydroelectricity, does hydrogen gas become a viable fuel, and only then if the storage problem can be economically overcome.

7.2.3 *Solar energy conversion*

Another way to generate hydrogen would be to use sunlight to "crack" water photochemically into its constituent gases. In a sense, this would be mimicking photosynthesis. The difference is that in the latter no hydrogen is produced; the reducing power being used to generate the biological reducing agent NADPH. Oxygen, of course, is produced and liberated by the organism into the atmosphere.

The last big oil crisis (during the 1970s) catalysed intense interest in solar energy conversion. The most successful systems to date are photovoltaic devices that convert solar energy directly into electricity. These systems can reach efficiencies of around 25–27%, and compare favourably with photosynthesis which, on a global scale, is about 1% efficient. This is a figure that represents the percentage of the total solar energy falling on the earth that plant photosynthesis converts to energy-yielding compounds. Certain plants are much more efficient than this, e.g. sugar cane can be up to 10% efficient.

One per cent efficiency, while being enough to support a global ecology, is not enough sadly to power an industrial or post-industrial society, even when we consider that our vital reserves of coal, oil, and gas are the result of past photosynthetic activity. In some countries, biomass is being turned into energy, e.g. alcohol from sugar cane in Brazil and diesel fuel from sunflower seeds in South Africa.[4]

Silicon photovoltaic cells, although more efficient, are expensive to produce and would have to cover large land masses in order to produce sufficient quantities of solar-powered electricity. For example, to replace the 2000 MW power station at Didcot, Oxfordshire in the UK with photovoltaic solar cells would require a land area of about 16 km^2.

Electrochemical splitting of water produces hydrogen but, as mentioned previously requires an externally supplied source of electricity. Reversible fuel cells are being developed that can be connected to mains electricity to generate hydrogen (which is stored as a metal hydride), and then used the stored hydrogen to produce electricity. But no matter how much cleaner and more efficient these fuel cells are, they still depend on fossil fuel-generated electricity for recharging.

Photoelectrochemical cells offer a possible solution to the recharging problem, and are beginning to challenge photovoltaic devices in terms of efficiency. They differ from electrochemical cells in that one electrode is made of a semiconductor (e.g. gallium arsenide or cadmium sulphide), which absorbs light energy and generates electrons which, on reaching the surface of the semiconductor, go into the solution where they are used to perform an electrochemical reaction, e.g. splitting of water. Over the last decade, not only have these cells begun to increase in efficiency, but they have also begun to be more reliable.[5]

7.2.4 Photochemical splitting of water

The photochemical splitting of water into hydrogen and oxygen has also

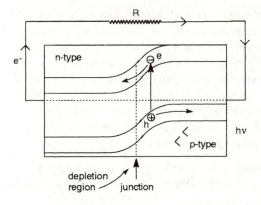

a Photoelectric cell based on a p-n junction

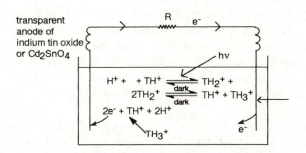

b Photoelectrochemical cell

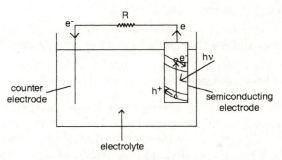

c Photogalvanic cell made of iron and thiomine

Fig. 7.7 Photoelectric, photoelectrochemical, and photogalvanic cells.

been studied intensely. The idea here has been that with the right mix of chemicals in water, solar energy could be utilised to generate hydrogen gas which, assuming it could be separated, could then be used as a clean-burning, pollution-free fuel.[6] Progress has been slow and beset by problems, so it will be some time before water photolysis can become a practical method of solar energy conversion. Porphyrins have played a part in helping to make laboratory experiments in this area work.

Systems that attempt hydrogen generation via the photochemical splitting of water usually consist of the following.

(i)   A photosensitiser to collect light energy and convert it into excited electrons (essentially performing the same function as chlorophyll in photosynthesis).

(ii) An electron relay to guide the excited electrons away from the photosensitiser (so that the electron does not combine with it, thus wasting the energy that has been collected).

(iii) A colloidal precious metal catalyst (usually platinum) to collect the electrons and use them to reduce water to hydrogen.

(iv) A sacrificial electron donor, whose job is to replenish the electrons lost by the photosensitiser: as it is oxidised so it is destroyed.

A similar system can be envisaged for the oxygen-generating side of photochemical water splitting. Here, instead of being photoreduced, water has to be photooxidised. The sensitiser now doubles as an electron relay. When it is photoexcited, it now donates its excited electron to a sacrificial electron acceptor, which acts as an electron sink and is gradually removed from the reaction. The oxidised photosensitiser then replenishes its electron shortage from the catalyst which, on being oxidised, takes electrons from water. Both the photoreduction and the photooxidation reactions will continue as long as the sacrificial electron donor and acceptor are present. However, it has not yet been possible to get both reactions to occur in each other's presence. Under those circumstances, the sacrificial donor and acceptor would be unnecessary as water would fulfil both functions, leading to a truly cyclic system.[6]

Compared with photosynthesis, which photochemical water splitting crudely mimics, the latter has all the elements of the system moving randomly in solution. Here, the various elements of the reaction are free to collide with each other and so lose their energy, or become involved in wasteful side reactions that end up consuming them. Photosynthetic systems, on the other hand, have all the active ingredients bound to cell membranes and positioned so that all the photophysical events (e.g. photon capture and electronic excitation) and their consequences (e.g. electron relay and water photooxidation) occur as quickly and as efficiently as possible. An

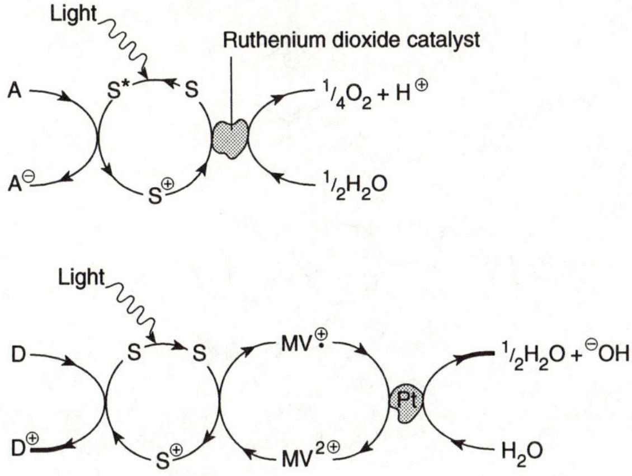

Left top: splitting water to make oxygen.
A = electron acceptor (a cobalt complex)
A^- = reduced oxygen acceptor
S = sensitiser, ruthenium (II) trisbipyridyl, doubling as electron-relay
S* = sensitiser excited by light
S^+ = oxidised sensitiser

Left below: Gratzel's hydrogen generating system.
D = electron donor (triethanolamine, cysteine or ethylene diamine tetraacetic acid
D^+ = oxidised electron donor
S = sensitiser, ruthenium (II) trisbipyridyl or zinc (II) tetramethylpyridyl
S* = sensitiser excited by light
S^+ = oxidised sensitiser
MV^{2+} = electron-relay (methyl viologen)
Pt = Platinum catalyst

Fig. 7.8 Systems for the photocatalytic oxidation and reduction of water.

example of how photochemical water splitting can go wrong (because the elements of the system are simply dissolved in water) is that the hydrogen produced can react with the electron relay (reducing it and so removing it from the system) and poison the catalyst. In both cases, hydrogen production stops. These problems are not insurmountable, however. The main problems with these systems have been with the photosensitisers.

In photosynthesis, the sensitiser chlorophyll's excited state survives for a mere fraction of a nanosecond before the excited electron is captured by quinone electron acceptors, which transfer the electron across a membrane and away to electron-transporting enzymes. In contrast, the photosensitisers used in photochemical water splitting have to survive in an excited state for up to several hundred microseconds (thousands of times longer) in a hostile environment, awaiting a favourable collision with the right molecule before giving up their energy. In that time, the energy, and indeed the sensitiser, can be degraded and lost. And that is where porphyrins come in.

Synthetic porphyrins have played an important part in the development of photochemical systems for solar energy conversion.[6,7] In particular,

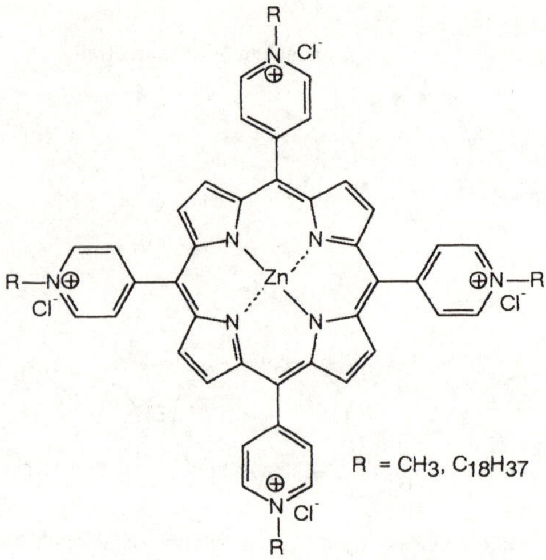

Fig. 7.9 Synthetic zinc porphyrins for solar energy conversion.

water-soluble zinc porphyrins have been used to photosensitise water reduction[6] and oxidation,[8] and they have also been shown to exhibit a photogalvanic effect.[9] (When light is shone into a solution containing the dissolved porphyrin and two electrodes connected to a battery, then a small current is passed.)

The reason why porphyrins have been used is because they have good light-absorbing properties and favourable redox potentials for photochemical water splitting. Also, with zinc occupying the central hole in the middle of the porphyrin, the excited state most favoured by the metalloporphyrin is the triplet, where the excited electron and the unpaired electron left behind have parallel spins. This excited triplet state has a long lifetime. However, in aqueous solutions these porphyrins, their excited states, and the important reactive intermediates derived from them are chronically unstable. For example, zinc porphyrins themselves are unstable to acids, the central metal being displaced by protons. And photoreduction of water to hydrogen is best done at a low pH because the redox potential for this reaction is lower. In other words, it is easier to do the reaction in acid solutions. Then, the excited triplet state can be destroyed by several competing molecular processes involving the interaction of the porphyrin with other porphyrin molecules, either in the ground state or the excited state, and with other ions.[10]

After the porphyrin has been excited, it can lose an electron to form a radical cation, or it can gain an electron to form a radical anion. Each of these species is highly reactive in its own right and they rapidly and irreversible disproportionate in water to give, respectively, dications and

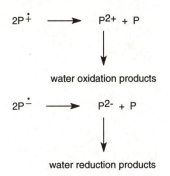

$$2P^{\overset{+}{\cdot}} \longrightarrow P^{2+} + P$$

water oxidation products

$$2P^{\overset{\cdot}{-}} \longrightarrow P^{2-} + P$$

water reduction products

Fig. 7.10 Disproportionation of zinc porphyrin π-cation and π-anion radicals in aqueous solution.

phlorin anions,[11] which are useless for hydrogen or oxygen generation and react further, ultimately destroying the porphyrin.

So, what does all this add up to? One of the best hydrogen-producing systems via photochemical water splitting was developed at the Royal Institution of Great Britain by Nobel Laureate, Lord Porter and his research group back in the early 1980s. They achieved a respectable light to hydrogen conversion efficiency of around 60%. Unfortunately, because of the instabilities of the metalloporphyrin photosensitiser (and other problems already mentioned, with the electron relay and the catalyst) the system's lifetime is only a few hours.[12a] How could the porphyrin sensitiser, at least, be made more stable?

The chemical answer to this is to identify the sites on the porphyrin molecule where reactivity occurs and to block them. These sites are the *meso*-carbons and the metal centre. Nature, of course, solved this problem eons ago by embedding the sensitive chlorophyll molecule in the protein–lipid matrix of the chloroplast membrane. Synthetically, this problem is solved by substituting water-solubilising groups in the *meso*-positions so that a molecular chain covers that position and the metal centre like an umbrella.

The results are quite satisfying—radical cations and anions generated from these porphyrins are much more stable. For example, the radical anions of these porphyrins have a half-life of 12 ms compared with 1–200 us for the porphyrin usually used. Also, when these porphyrins were used for hydrogen generation, although the light to hydrogen conversion efficiency at 32% was not as good as Porter's system, the total yield of hydrogen was higher because the system produces hydrogen for a longer period of time— in fact, until the sacrificial electron donor is exhausted.[12b] At the moment, oxygen cannot be produced from the photooxidation of water using these porphyrins. If it could, a totally cyclic process would be possible.

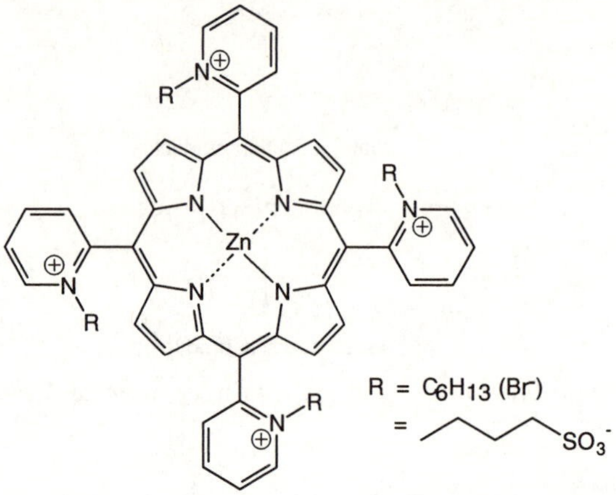

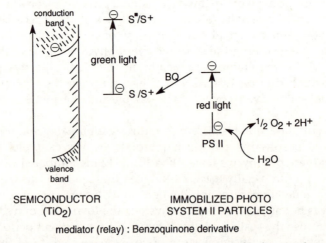

Fig. 7.11 Sterically hindered zinc porphyrins for solar energy conversion.

More recently, water photolysis has been used to generate electric current from the oxidation of water. Here, the shortcomings of the earlier systems were surmounted by using whole green plant PS II particles deposited on to photoelectrodes of polycrystalline titanium dioxide, derivatised with a dye.[12c] The photon to current conversion efficiency was observed to be 12%, while photocurrents of up to 35 Å cm^{-2} were obtained on illumination with white light of 35 mW cm^{-2} intensity. This system represents one of the first bioelectronic devices that manages to couple together a biological electron-transport chain with a semiconducting device.

Fig. 7.12 Schematic illustration of the photoredox events leading to the generation of electrical current via two-photon excitation of the dye-derivatised TiO$_2$ electrodes coated with PSII particles. [After Rao and Hall *et al.*]

7.2.5 *Photosynthetic model compounds*

Even if the photochemical splitting of water does not become a viable energy source for the future (and the chances of that happening are still very slim), the construction of model systems using porphyrin sensitisers has taught chemists much about the natural photosynthetic system. This is because the most significant events in photosynthesis are the first few fractions of a nanosecond after chlorophyll absorbs light energy, when photoinduced charge separation occurs. This has been shown to depend not only on the distances between the electron donor and acceptor, but also on their orientation to each other. So, as we saw in Chapter 3 during bacterial photosynthesis,[13] two bacteriochlorophyll b (BChl b) molecules are closely associated both spatially and electronically, to form a special pair electron donor known as P865 (see Figure 3.34). Another pair of BChl b molecules are placed edge to edge, relative to P865, and these in turn are adjacent to a pair of bacteriopheophorbide a (BPheo a) molecules. A pair of quinone electron acceptor molecules complete this picture.

Using information like this, it may yet be possible to design fully functioning laboratory models that convert solar energy into chemical potential. So far, many models have been prepared that, to a greater or lesser extent, demonstrate the dependencies of electron transfer rate constants on the distance between the donor and the acceptor, their orientation to each other, the free energy involved in the transfer, and the electronic coupling between them. Only a small number of these model systems actually possess the structural constraints to control both the distance and the orientation between the electron donor and acceptor. Some of these are shown in Figure 7.13.

Porphyrins with quinones attached (Figure 7.13) show a marked decrease in their singlet excited state lifetimes; for example, from nanoseconds to picoseconds. The relative fluorescence quantum yield (i.e. the ratio of the number of molecules fluorescing to the number of light quanta absorbed—a difficult number to obtain absolutely) also drops, from 0.1 in the case of an unquenched *meso*-tetraarylporphyrin, to 10^{-5} for the porphyrin tetraquinone in which the quinones are directly bonded to the macrocyclic *meso*-positions.[14] As soon as the electron is excited, it is transferred to the quinone electron acceptor. This effectively depopulates the porphyrin singlet excited state, which is observed as a reduction in intensity and a blue shift of the porphyrin B band. Subsequently many workers have expanded on this theme.[7]

Another interesting molecule is the triad of carotenoid, porphyrin, and quinone all linked together by covalent bonds (Figure 7.13). Excitation of the porphyrin transfers an electron to the quinone, making the latter negatively charged. The electron deficiency on the porphyrin is then satisfied by

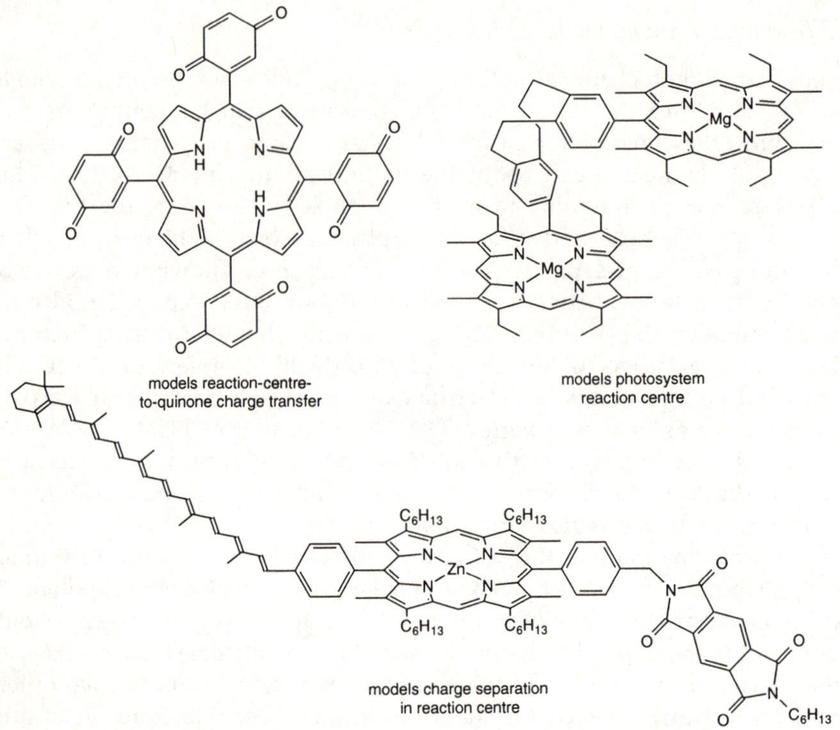

Fig. 7.13 A selection of porphyrin-based models for various processes during the light reaction of photosynthesis.

a donation from the carotenoid before the electron can be reclaimed from the quinone. The net effect is transfer of an electron from the carotenoid to the quinone, modelling the antenna-trapping activity of chlorophyll and carotenoid ancilliary pigments of photosynthesis.

Many of these model systems suffer from a fundamental problem, besides the purely synthetic one of putting them together. Namely, how to predict the rates of electron transfer in the model. The theory of electron transfer in molecules has been thoroughly worked out by R.A. Marcus, who received the 1992 Nobel Prize in chemistry for his labours. But even model systems that are closely related chemically can show huge variations in rates of electron transfer for no apparent reason. It is also certain that in these model systems, the solvent plays a significant part in determining the rates and energetics of electron transfer. So, even after a huge synthetic effort on the part of chemists, it is still not possible to make useful models that can be used to generalise the main features of photoinduced electron transfer.

In an attempt to overcome these limitations of model systems, Jeremy Saunders and his team at Cambridge University, have used coordination chemistry instead of covalent chemistry to increase the synthetic versatility and to maintain conformational control.[15] They use a zinc metalloporphyrin electron donor and an amine acceptor linked to a "spacer" molecule that coordinates to the zinc cation. Saunders has found that the coordinated acceptor fully quenches the porphyrin singlet excited state. This, in itself, is not remarkable, but then Saunders found that his donor–spacer–acceptor system was capable of discriminating between *meso*- and β-substituted porphyrins—the latter being about five times faster at quenching the porphyrin singlet excited state than the former. Also, rates of electron transfer and back reaction were similar, regardless of whether the spacer group was aliphatic or aromatic. This indicates that an aromatic pathway does not necessarily speed up electron transfer, and suggests that "through space" electron transfer might also be important.

7.3 Porphyrins and cancer therapy[16]

7.3.1 Introduction: HpD and PDT

The use of photosensitisers to bring about photoreactions in biological systems dates back to the late nineteenth century, when Finsen discovered that the skin condition *Lupus vulgaris* could be treated using ultraviolet light. Princess Alexandra, the future queen of King Edward VII and a great patroness of the medical sciences, brought the discovery to the London Hospital. In 1913, Meyer-Betz demonstrated the phototoxicity of porphyrins by injecting himself with 200 mg of "hematoporphyrin derivative" (HpD—more of this substance later). By 1924, Policard had discovered that certain malignant tumours accumulated porphyrins. It wasn't until 1975, however, that such a fortuitous combination was found to be beneficial in detecting and treating cancers. By 1976, the first successful trials with human volunteers in photodynamic therapy, or PDT, had been initiated.

HpD has been the most commonly used PDT agent and is derived synthetically from hematoporphyrin (Figure 7.14) by treating it with acetic and sulphuric acids. This produces a mixture (called Stage 1) of 3,8-disubstituted deuteroporphyrins in which the substituents consist of various combinations of vinyl, 2-acetoxyethyl, and 2-hydroxyethyl groups. The main component of this Stage 1 HpD mixture is the diacetoxy derivative (Figure 7.14). In order to get this mixture into a condition in which it can be administered (called HpD Stage II), it is dissolved in alkali. This caused hydrolysis or elimination of the acetoxy groups to give a complex "witches brew" of monomeric, dimeric, and oligomeric porphyrins. Only a few of the compounds in this mixture are known to be PDT-active *in vivo*, and these

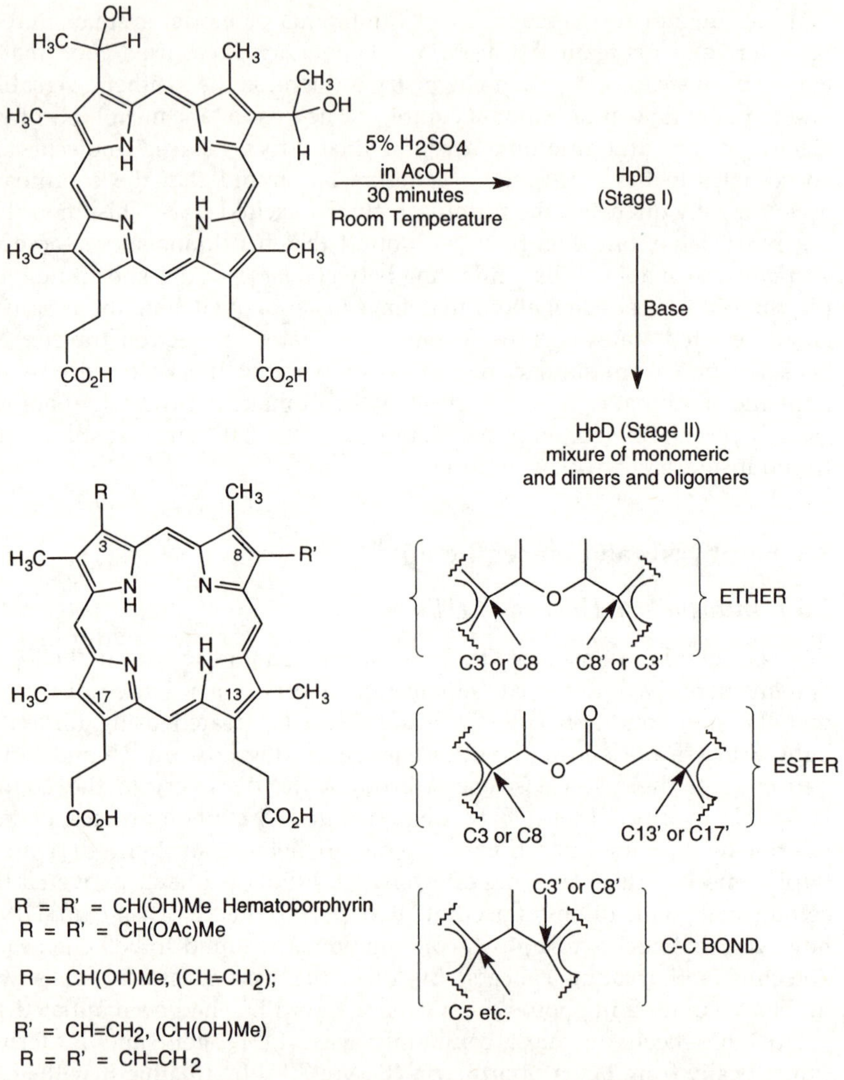

Structures of hematoporphyrin and its monomeric and oligomeric congeners.
The righthand side of the figure shows postulated interporphyrin linkages

Fig. 7.14 HpD synthesis and some of its components.

are dimers and oligomers (Figure 7.14); but it is not yet clear which. Even a
partially purified HpD mixture, called photofrin II and used in studies to
determine the active constituents of HpD, contains porphyrin monomers,
ester-linked and ether-linked dimers, and oligomers of up to six porphyrins,
with β-substituents consisting of hydroxyethyl and vinyl groups.

HpD has been used to treat a variety of tumours and bladder cancers, breast cancers, and certain ocular cancers. Just as bilirubin photosensitises the production of singlet oxygen to bring about its own destruction in the photodynamic treatment of neonatal hyperbilirubinaemia (see Chapter 6), so components of HpD also produce singlet oxygen. And because the HpD tends to localise in tumour cells, so photosensitisation leads to their destruction.

7.3.2 Seven golden rules for a good PDT agent

There are, however, many reasons why HpD is not the most ideal photosensitiser for the photodynamic treatment of cancers. For instance, it is a mixture of compounds, some of which are PDT-inactive. Consequently, it is difficult to reproduce similar photodynamic properties from different HpD samples. HpD also localises, to a lesser extent, in healthy tissues, leading to photosensitivity of the patient long after injection of HpD (about 1 month). Perhaps the biggest disadvantage of HpD though, is that it absorbs light in the wrong region of the spectrum for PDT. Let us see what this means.

The porphyrins present in HpD have their maximum absorption around 400 nm (the B band or Soret region), in the near-UV. However, during PDT, the light that is to excite the porphyrin has to pass through tissue. This has the effect of absorbing the light in the Soret region, but attenuates to a much lesser degree, light in the red end of the spectrum, around 620–630 nm (hold your hand up to a bright light source: the light that penetrates near the edges of your fingers is bright red). Unfortunately, it is in this region that HpD has its smallest absorption band. This means that HpD is inefficient at light absorption, and therefore produces only a small photodynamic effect. For several years now, the hunt has been on for more efficient photosensitisers of PDT—ones that absorb more powerfully in the red end of the spectrum. The following rules for what constitutes a good PDT photosensitiser have been enunciated.

1. It must be a pure compound with a reproducible synthesis.
2. It must be activated at wavelengths >650 nm to ensure better absorption of tissue-penetrating red light, and so sensitisation by an external light source.
3. It must be non-toxic in the absence of light.
4. Its excited states (particularly the triplet excited state) must be long-lived enough to enable it to photosensitise the production of singlet oxygen (which is believed to be the photoreagent responsible for the PDT effect).
5. It must localise specifically in the tumour.
6. It must clear rapidly from the body after it has done its work.

7. It must be soluble in the body's tissue fluids so that it can be injected and carried round the body to the tumour site.

7.3.3 *Second-generation PDT agents*

Many of the properties listed above are being incorporated into new, second-generation porphyrin photosensitisers and into new porphyrin delivery systems. For example, several classes of compounds have been synthesised based on the structure of *meso*-tetrakistetraphenylporphyrin

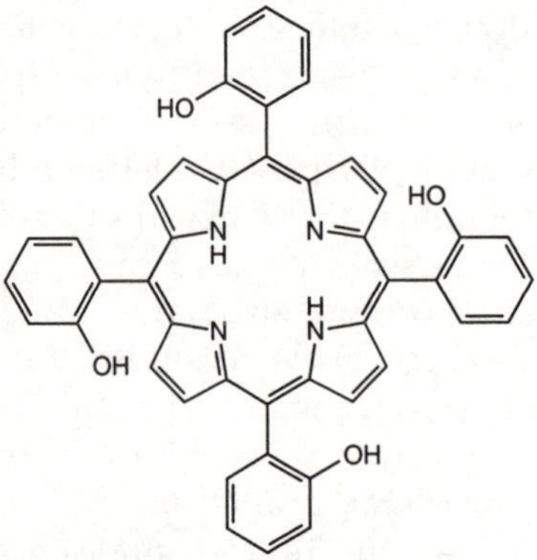

5,10,15,20-Tetra(*o*-hydroxyphenyl)porphyrin
o-THPP

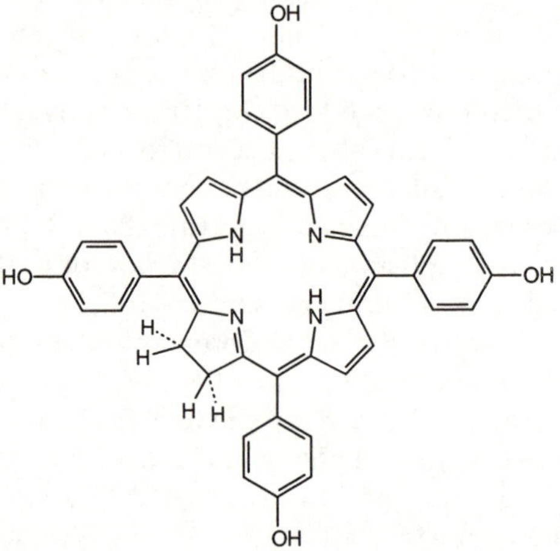

5,10,15,20-Tetra(*p*-hydroxyphenyl)chlorin
p-THPC

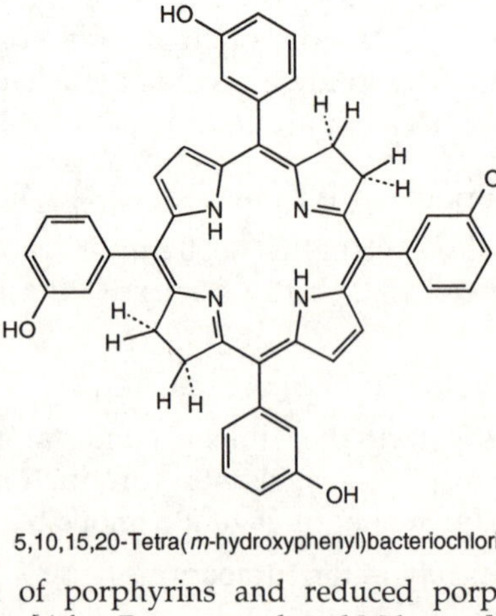

5,10,15,20-Tetra(*m*-hydroxyphenyl)bacteriochlorin

Fig. 7.15 A selection of porphyrins and reduced porphyrins in the *meso*-tetra (hydroxyphenyl) series. [After Bonnett *et al.* and Milgrom.]

(TPP), such as the hydroxyphenyl porphyrins and chlorins. These have stronger absorption at longer wavelengths (643 nm for the porphyrin and 715 nm for the chlorin) than HpD or photofrin II and are more water soluble because of their hydroxy groups. Chlorins are particularly promising PDT agents precisely because they have much stronger absorption in the red end of the visible spectrum. Figure 7.16 shows some of the chlorin derivatives that are being experimented with.

Of the seven guiding principles enunciated above, the one that has taxed the skills of chemists most has been to ensure that the chromophore has the desired long wavelength absorption. Reduction of one of the macrocycle double bonds to produce chlorins, as above, is one successful strategy. Another is to expand the macrocycle. Thus, phthalocyanines (Pcs) have been intensively investigated. Typically, they absorb in the range 675–700 nm and introduction of sulphonate groups gives them water solubility (Figure 7.17). In addition, chelation of a diamagnetic metal (e.g. aluminium or zinc) by the phthalocyanine, enhances phototoxicity; the Pc being non-toxic in the absence of light. Other advantages of metallo-Pcs is good retention by tumours, good PDT activity, and low skin photosensitivity.

Macrocycle expansion is an interesting challenge to the synthetic chemist and they have not been slow to rise to the occasion. Thus, expanded porphyrins have been prepared (Figure 7.18), such as the sapphyrins, pentaphyrins, porphycenes, texaphyrins, and porphyrin vinylogues. Apart from being interesting theoretically (they pose a challenge to the validity of Hückel's $[4n+2]$ rule for aromaticity, as some of these compounds, e.g. pentaphyrins, have 22 π-electrons), they all absorb strongly at long wavelengths and they show considerable promise as PDT agents.

One of the main problems with many of the photosensitisers mentioned above is that their specificity for tumour cells over healthy cells is not great enough. Even in a promising new family of second-generation photosensitisers, such as the water-soluble phthalocyanines, the specificity for tumour cells over healthy cells is only about 4:1. This means that under illumination, healthy tissue will experience some photodamage.

One new approach that is being developed to enhance tumour specificity used monoclonal antibodies (mAb), raised against a particular type of tumour, as a kind of molecular "guided missile", with a porphyrin attached to the mAb as the light-activated "warhead." Tumour cells have different surface antigens to normal cells, and it is possible to raise mAbs specific for these antigens. Tumour cells also differ from normal cells in that they express a large number of low-density lipoprotein receptors. Hydrophobic photosensitisers may then be incorporated into a lipid moiety of such a receptor, and taken into the tumour cell via lysosomes.

As mentioned earlier, the mechanism of PDT action (via singlet oxygen production) is similar to the way bilirubin photosensitises its own destruction

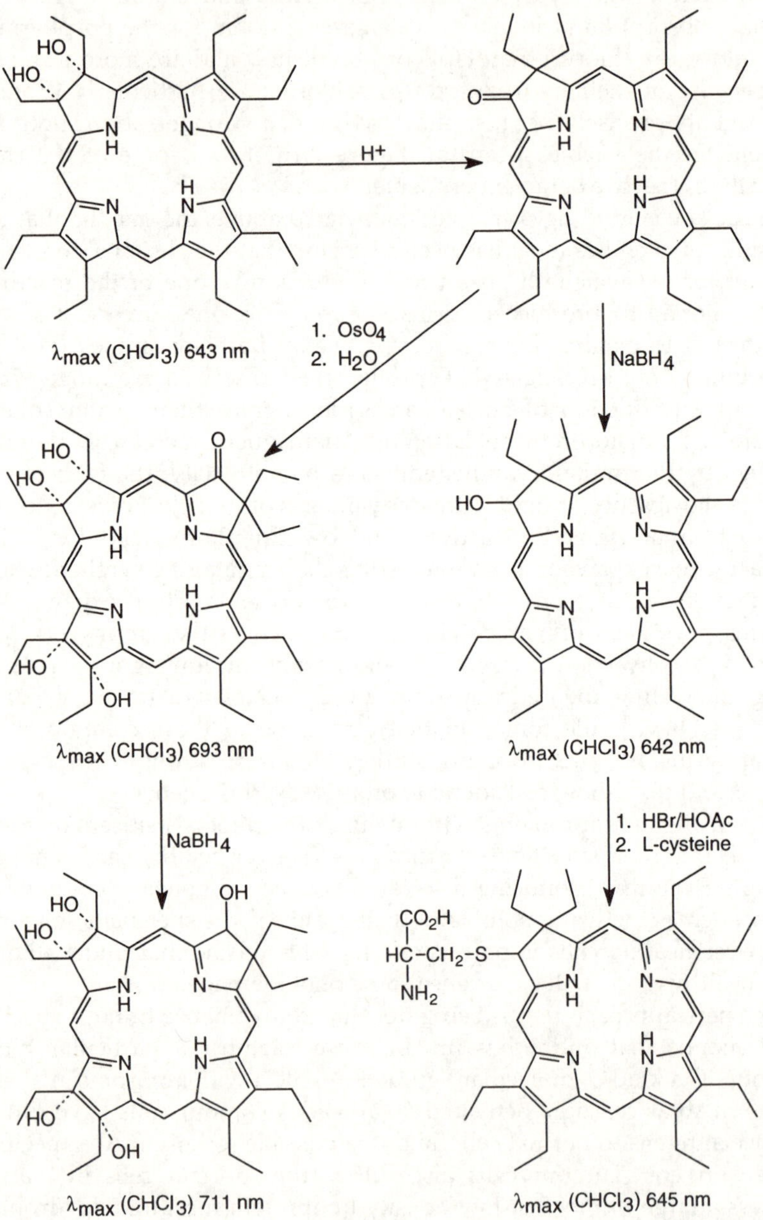

Fig. 7.16 Preparation of β-hydroxy and β-keto-substituted chlorins for PDT (after Bonnett and Berenbaum).

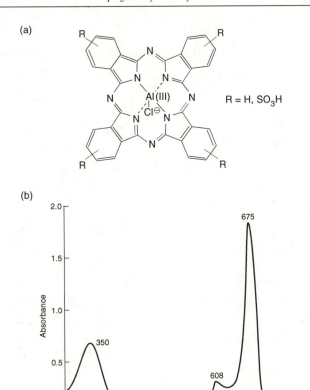

Fig. 7.17 (a) Chloroaluminium tetrasulphophthalocyanine. The structure illustrates the general formula for mono-, di-, tri-, and tetra-sulphonated phthalocyanines. (b) The UV–visible spectrum.

in the treatment of neonatal hyperbilirubinaemia. Thus, excitation of the porphyrin leads to a π–π^* transition of an electron from a HOMO to a LUMO, leading to a singlet excited state. This can fluoresce to the ground state, in which case the fluorescence can be used as a diagnostic tool for the detection of cancer cells. The singlet excited state can also convert to a triplet state, in which the electrons in the HOMO and LUMO have parallel spins. It is in this state that the excited porphyrin will interact with ground state oxygen (also a triplet, $^3O_2$), the former being quenched, while the latter is converted to singlet excited state oxygen, $^1O_2$. The singlet oxygen that then attacks sensitive cell components such as membranes and nucleic acids.

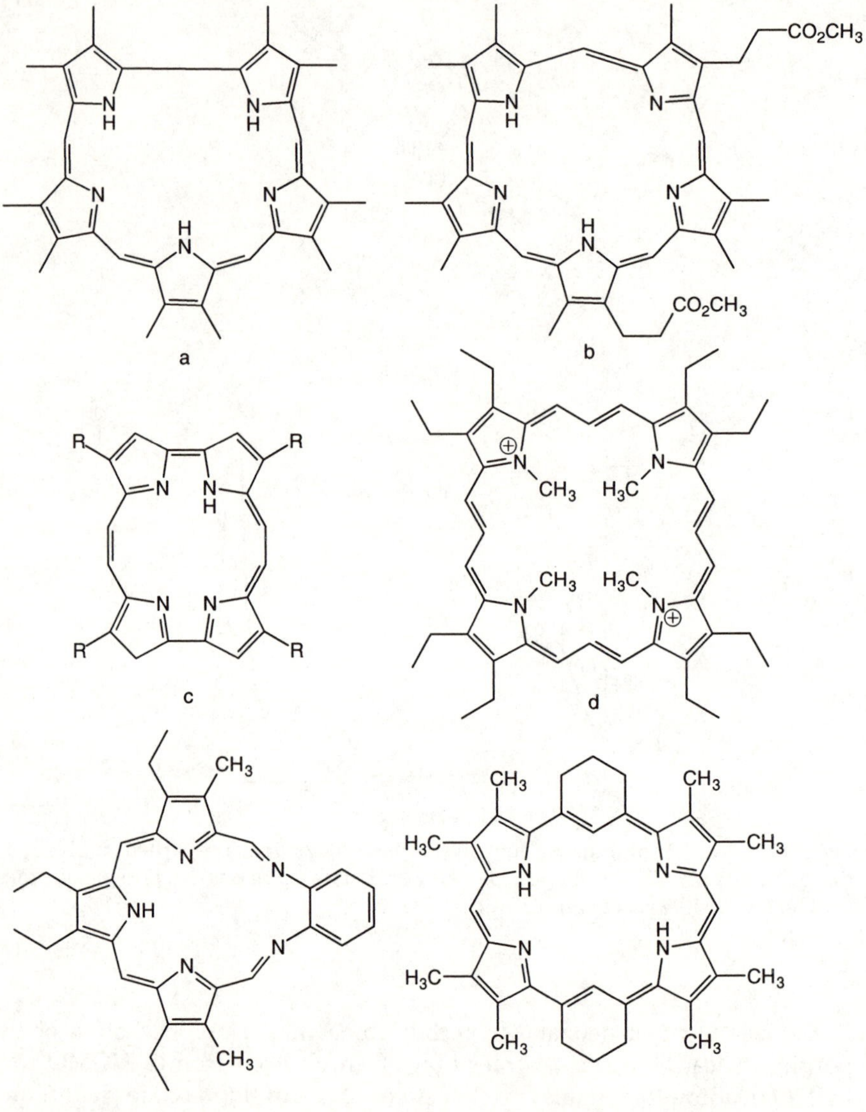

Fig. 7.18 A selection of 'expanded' porphyrins for PDT: (a) sapphyrin, (b) penta-phyrin, (d) porphyrin vinylogue (fourfold expansion), (e) texaphyrin, (f) platyrin. [After Kreimer-Birmbaum and Sessler *et al.*]

7.3.4 *Porphyrins and DNA cleavage*

Certain porphyrins have also been shown to bring about cleavage of DNA, both chemically and photochemically, so that they have potential as antiviral agents. Cationic porphyrins (see Figure 7.19) are particularly good at cleaving DNA even though they are just as efficient at producing singlet oxygen photochemically as anionic porphyrins, which are not good cleavers.

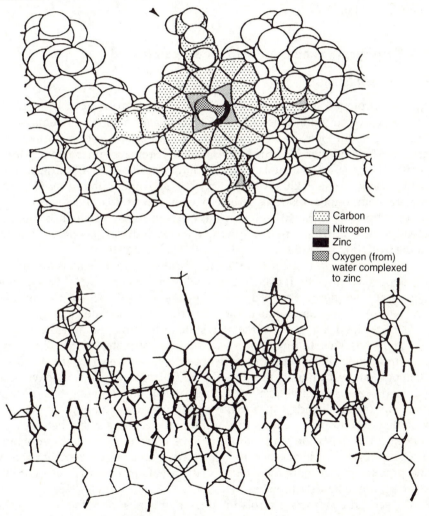

Carbon
Nitrogen
Zinc
Oxygen (from) water complexed to zinc

Fig. 7.19 Qualitative molecular modelling of the face-on interaction of ZnTMPyP (4) with the major groove. The expanded view (top) reveals the lack of interaction of one of the *N*-methylpyridinium groups with either of the sugar–phosphate chains.

The porphyrins appear to bind to DNA in one of three ways: (a) by inter-calation between pairs of nucleotide bases, usually guanine and cytosine, and they cause unwinding of a DNA supercoil; (b) by stacking on the outside of the DNA coil, on the sugar–phosphate backbone where there are adenine and thymine base pairs on the inside (this can cause aggegation of DNA strands if the porphyrins stack with themselves); and (c) by again binding to the outside but opposite guanine and cytosine as well as adenine and thymine base pairs. However, it is by no means clear just how por-phyrins bind to DNA.

7.4 Porphyrins as hemoprotein model compounds

7.4.1 Introduction

In hemoproteins, iron [as Fe(II) or Fe(III)] porphyrins act as prosthetic groups, carrying out a wide range of biochemical functions. In hemoglobin and myoglobin, for example, oxygen is, respectively, transported and stored. The oxidation state of the Fe(II) cation remains unchanged. Mean-while, peroxidases and catalases mediate the metabolism of hydrogen per-oxide, cytochrome P450 oversees oxygen activation and insertion into organic compounds, oxidases reduce oxygen to water, and cytochromes b_5 and c help to pass electrons down the electron-transport chain in cellular metabolism. All these processes involve controlled and reversible changes in the oxidation state of the central coordinated iron cation.[17]

The difference in heme activity in all of these hemoproteins is a result of the apoprotein moiety. It exercises control over the heme by (a) varying the nature and strength of the iron axial ligands and (b) varying the number and nature of the amino acids in the iron cation's immediate vicinity.

The first condition governs the spin state of the iron and access to one of its axial sites. This, in turn, controls the redox chemistry of the iron. Thus, one imidazole ligand in the fifth position and oxygen in the sixth (hemoglobin and myoglobin), puts the Fe(II) cation in a less reactive, dia-magnetic low-spin state, ideal for oxygen transport. Changing one of these axial imidazole ligands, e.g. for the sulphur of a methionine amino acid residue (cytochrome c), allows the Fe cation to cycle reversibly between the +II and +III oxidation states, and prevents oxygen binding on reduction. The sulphur probably also acts to help convey electrons from other parts of the protein (cytochrome c usually works in conjunction with cytochrome oxidase) to the Fe(III) cation.

The second condition controls the degree of exposure of the heme to the hemoprotein's external environment and access of ligands and substrates to its active site. Thus, heme buried deep within the hydrophobic interior of a globin apoprotein is ideally suited to the non-oxidative binding of oxygen.

The exclusion of water stops competition for the heme Fe(II) which, in the presence of water would be high spin and easily oxidised to Fe(III). Heme sited near the hydrophilic exterior of an apoprotein is well placed to cycle through a range of oxidation states, modulated by the porphyrin macrocycle. Clearly, nature has evolved a highly sophisticated set of supramolecular parameters which determine thermodynamic and kinetic criteria for controlling heme reactivity.

The purpose of synthesising hemoprotein model compounds is twofold. First, it gives an understanding into the nature and operation of those criteria in natural systems.[18] Secondly, by constructing models that mimic the supramolecular ability of enzymes for binding, recognition, and catalysis,[19a] the models themselves may find practical application, e.g. in he oxidation of organic compounds.[19b]

7.4.2 Supramolecular effects in hemoprotein model compounds

Early studies using porphyrin model compounds attempted to mimic reversible oxygen binding in myoglobin. This involved the synthesis of porphyrins with a variety of "straps", "caps", "tail bases", "picket fences" and "basket handles" that served to provide a degree of supramolecular control over the way oxygen bound to the iron cation.[20]

These compounds were moderately successful at reversibly binding oxygen at ambient temperatures, but only in aprotic solvents. Ether- and amide-linked "basket handle" hemoprotein models have been more extensively studied using electrochemistry to see how the superstructure around the iron porphyrin affects their reactivity.[18] Thus, hydrophobic ether linkages protect the cationic iron reaction centre from external solvation, so that reactions are more difficult (cf. myoglobin). On the other hand, the more hydrophilic amide linkages "locally solvate" the iron centre (by interaction between –NHCO– dipoles and build-up of negative charge on the central iron as it is reduced). This resembles ordinary solvation in terms of interaction energies but differs on entropic grounds—local solvation has a tiny reaction entropy compared to solution entropy. In a similar manner, the amide groups also affect the binding of axial ligands. Thus, the building of an ordered structure about the reaction centre (in this case, an iron atom) precisely controls its reactivity, just as in the natural hemoproteins.

A different kind of modelling of enzyme binding sites and processes has been studied by Jeremy Sanders and his team at Cambridge, UK. In one model, porphyrin face-to-face dimers and 1,4-diazabicyclo[2,2,2]octane (DABCO) (in conjunction with proton NMR) were employed as a simple two-component system in which the distance between the two porphyrin moieties could be varied. Depending on this distance, large variations were observed in the way DABCO was bound, even though the differences in structure of two co-facial dimers was relatively small.[21a] Furthermore, by studying the kinetics and thermodynamics of DABCO binding, the Cambridge team were able to estimate the strength of the $\pi-\pi$ interaction between the porphyrin moieties. This approach was extended to more flexible porphyrin dimers that could accommodate larger ligands within their cavities,[21b] and led to a simple electrostatic model that explains the geometry and theoretical basis of such $\pi-\pi$ interactions, even predicting when they should occur.[21c]

Sanders went on to synthesise a second generation of porphyrin hosts using guest amine-type molecules as templates. This enabled cyclic porphyrin dimers and trimers to be prepared.[22a] The trimer recognises and binds large polyoxometalate anionic clusters (e.g. $PW_{12}O_{40}{}^{3-}$) and anionic metal–carbonyl clusters containing osmium and carbon (e.g. $Os_{10}C(CO)_{24}{}^{2-}$).[22b]

The thrust of the Cambridge group's work has been to construct model enzyme systems capable of binding, recognition, and catalysis. Such systems should be easy to construct from readily available, large and rigid building blocks and have built-in moieties that make it easy to probe the geometry and binding by spectroscopy. Furthermore, there should be convergent binding sites for two or more substrates.[19a] To this end, they synthesised a double steroid-capped porphyrin.[23] The zinc compound selectively bound a variety of simple amines; in particular purine, which is capable of H-binding to the OH groups of the steroidal caps. A ligand-tailed version of this metalloporphyrin was also prepared by acylating one of the steroidal OH groups with 3-carboxypyridine. Interestingly, no reaction occurred with the zinc-free porphyrin indicating that the zinc helps to position the pyridine for rapid, intramolecular reaction.

7.4.3 Models of cytochrome P450

As mentioned in Chapter 4, cytochromes P450 are membrane-bound monooxygenase enzymes that homogeneously catalyse oxygen atom transfer to bound non-polar substrates.[24] In fact, enzymes collectively classed as P450 are not really cytochromes: their function is oxygen-atom transfer, not electron transport.

P450s use NADPH to reduce dioxygen; one O atom is reduced to water while the second is transferred to a wide variety of endogenous (e.g. steroids, fatty acids, leukotrienes and prostaglandins) and exogenous (e.g.

Fig. 7.20 A variety of 'strapped' and 'capped' hemoprotein model porphyrins. [From D. Lexa and J.M. Saveart, *Proceedings of the International Symposium on Redox Mechanisms and the Importance of Interfacial Properties in Molecular Biology*, 3rd Meeting, G. Dryhurst and N. Katsumi (eds), Plenum, New York (1988).

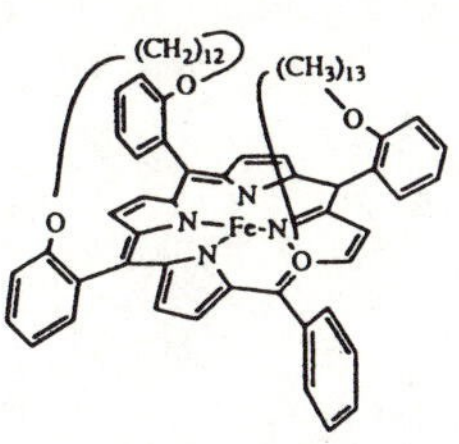

$(O(CH_2)_{12}O)_2$

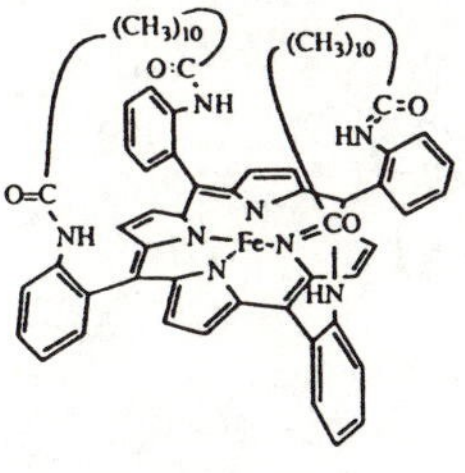

$(NHCO(CH_2)_{10}CONH)_2$

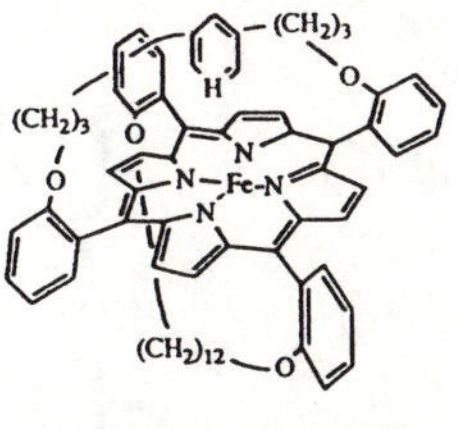

$(O(CH_2)_3(3-Py-5)(CH_2)_3O)(O(CH_2)_{12}O))$

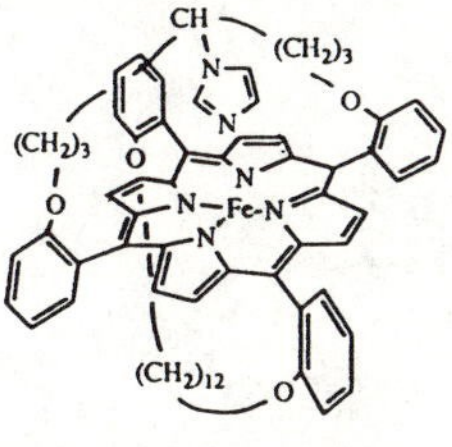

$(O(CH_2)_{12}O)_2$

$(NHCO(CH_2)_{10}CONH)_2$

$(O(CH_2)_5CH(1-Im)(CH_2)_5O)(O(CH_2)_{12}O))$

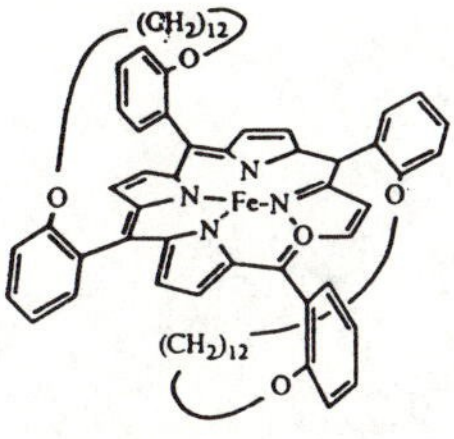

$(O(CH_2)_{12}O)_2$

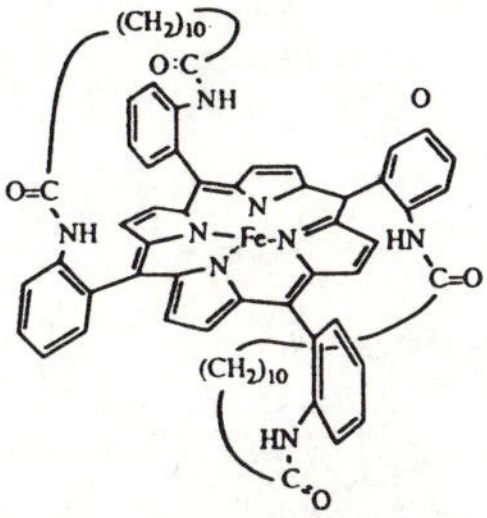

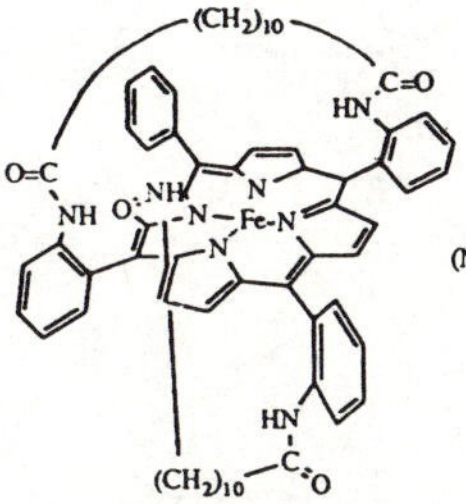

$(NHCO(CH_2)_{10}CONH)_2$

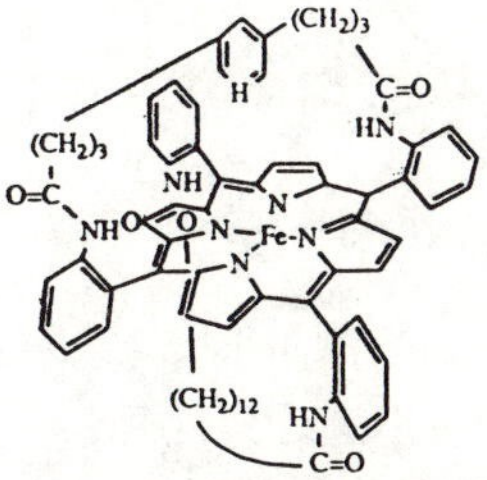

$(NHCO(CH_2)_2(3-Py-5)(CH_2)_2CONH)(NHCO(CH_2)_{11}CONH)$

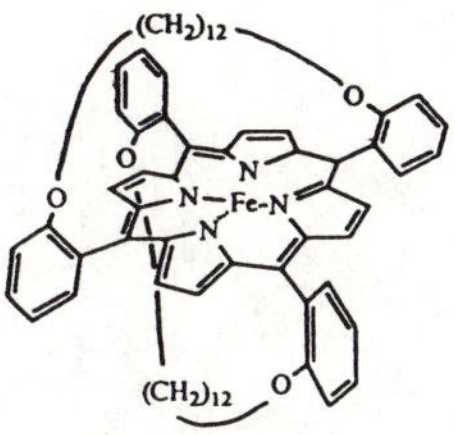

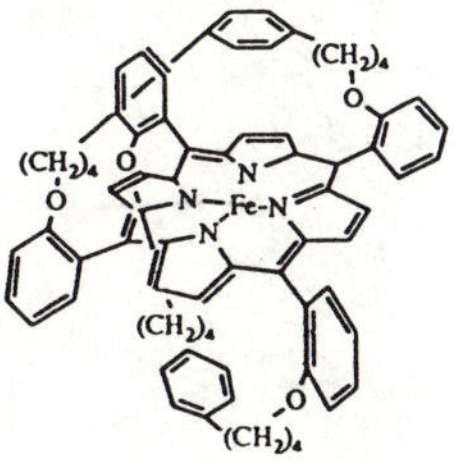

$(O(CH_2)_4Ph(CH_2)_4O)_2$

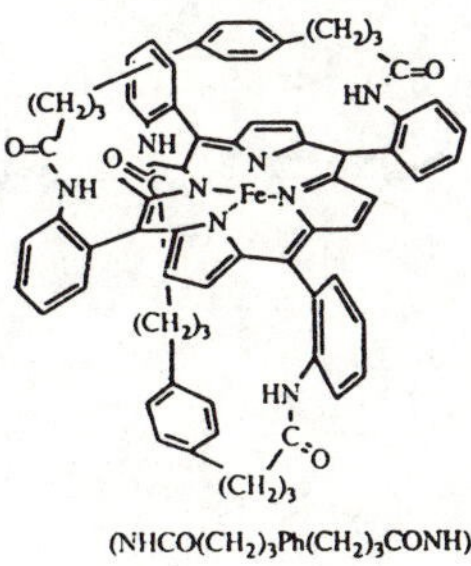

$(NHCO(CH_2)_3Ph(CH_2)_3CONH)_2$

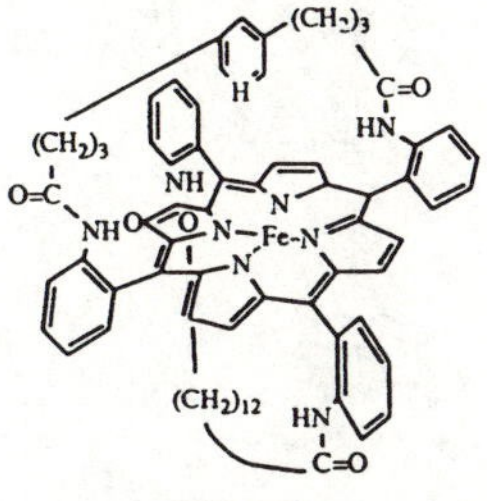

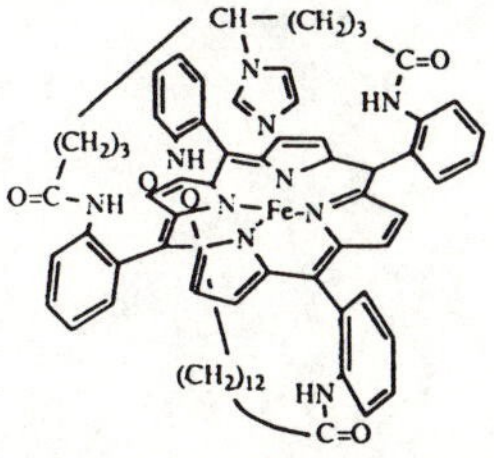

$(NHCO(CH_2)_3(3-Py-5)(CH_2)_3CONH)(NHCO(CH_2)_{10}CONH)$

$(NHCO(CH_2)_4CH(1-Im)(CH_2)_4CONH)(NHCO(CH_2)_{10}CONH)$

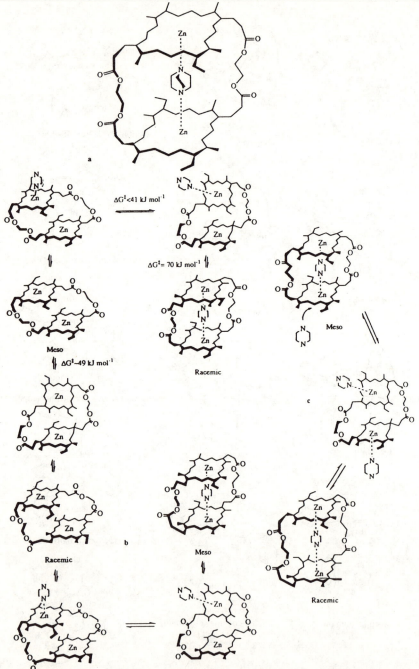

Fig. 7.21 (a) A porphyrin–dimer–DABCO complex and exchange processes in the presence of (b) excess porphyrin dimer and (c) excess DABCO. [After Hunter *et al.*]

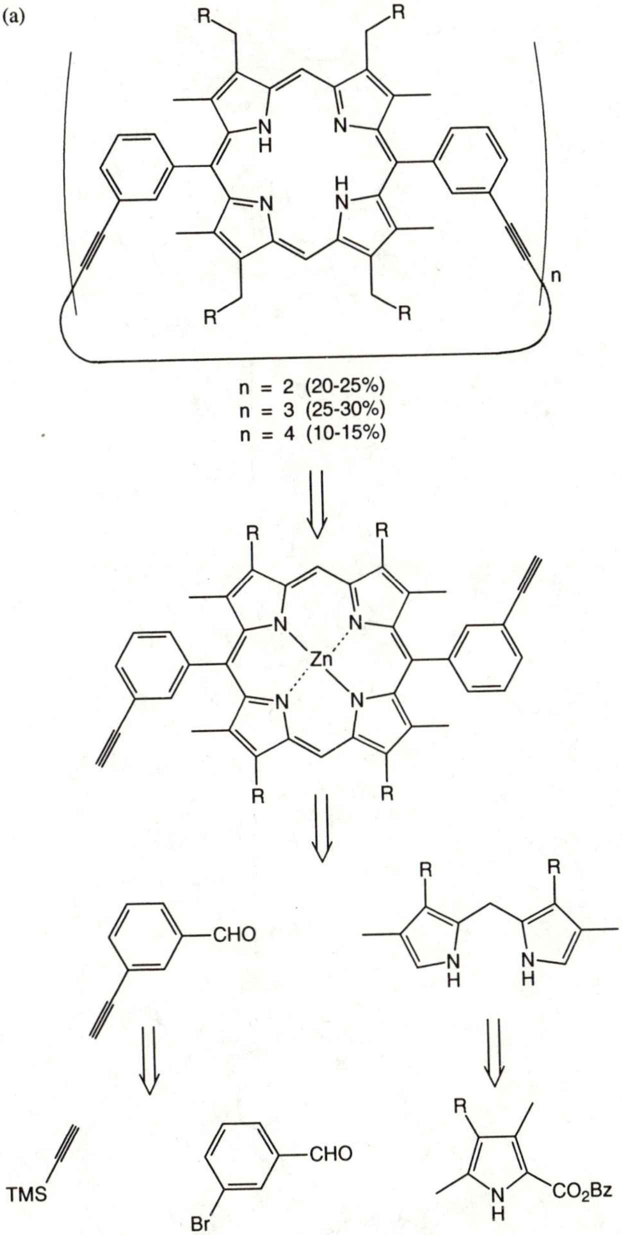

Fig. 7.22 (a) Retrosynthetic synthesis of porphyrin oligomers without amine templates. [After Sanders *et al.*] (b) Synthesis of porphyrin oligomers with amine templates. [After Sanders *et al.*] (c) Synthesis of oligomeric porphyrins binding a metal–carbonyl cluster and a chiral aluminium template.

(b)

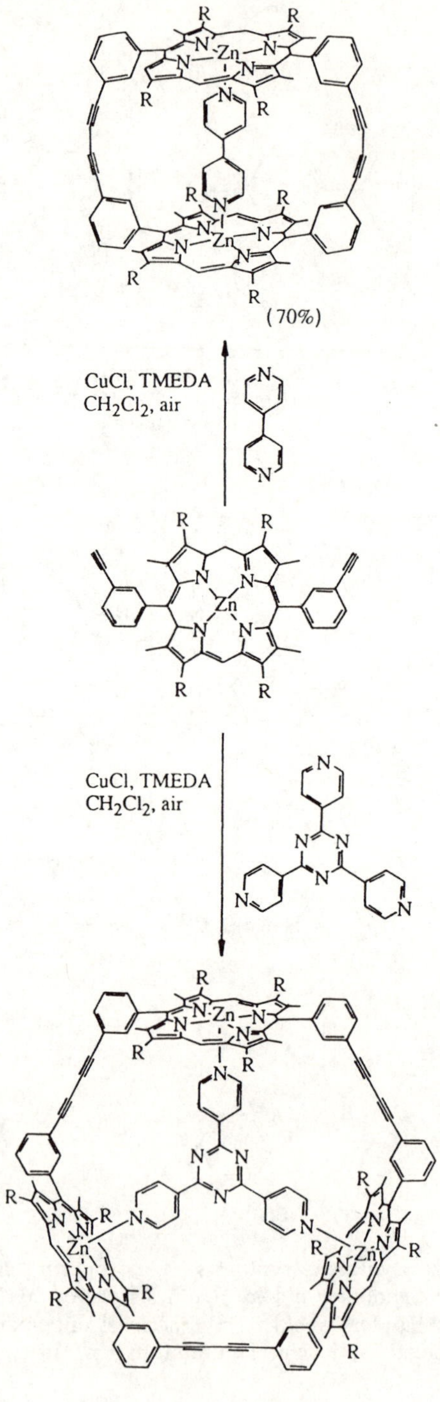

(70%)

CuCl, TMEDA
CH$_2$Cl$_2$, air

CuCl, TMEDA
CH$_2$Cl$_2$, air

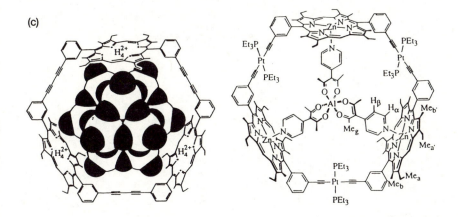

drugs, pesticides, anaesthetics, solvents, and chemical carcinogens) sub-strates. The structure of the camphor-hydroxylating enzyme, for example, from *Pseudomonas putida*[25] shows that the camphor-binding site, the heme, and the oxygen-binding site are all in close proximity within the protein. In addition, the camphor-binding site is lined with hydrophobic amino acid residues and protected from the outside world by a reversible cap. The heme is attached to the protein only via the axial sulphur atom of a cysteine residue, with a water molecule (in the substrate-free enzyme) near enough to the iron to be a second axial ligand. Thus, according to the criteria men-tioned earlier, the heme-bound iron cycles through a range of oxidation states relatively easily (see Figure 4.33).

In its resting state, the iron complex is in equilibrium between a five-coor-dinate, high-spin Fe(III) and a six-coordinate, low-spin Fe(III) complex. Bind-ing of the substrate favours the five-coordinate species as the water molecule in the sixth position is displaced. Uptake of an electron is favoured by the now high-spin Fe(III) which is reduced to Fe(II) and binds oxygen to form a stable, low-spin, six-coordinate intermediate. On uptake of a second electron, the oxygen undergoes cleavage to yield water and an oxo–iron intermediate. It is this that oxygenates the substrate. The thiolate fifth ligand pushes electron density onto the coordinated oxygen, bringing about cleavage.

Molecules containing reduced oxygen species, e.g. alkyl hydroperoxides, peracids, periodate, hypochlorite, hydrogen peroxide, amine oxides, and iodosoarenes, are also capable of reducing dioxygen in the presence of P450. It is this that gave the impetus to the development of models based on simple porphyrin systems.

Most of the metalloporphyrin systems used as P450 models are variations on the theme of *meso* tetraarylporphyrins,[24] with a variety of substituents in the *meso*-positions. These models manage to combine effective stereospecific oxidation of the products, protection of the iron porphyrin catalyst from

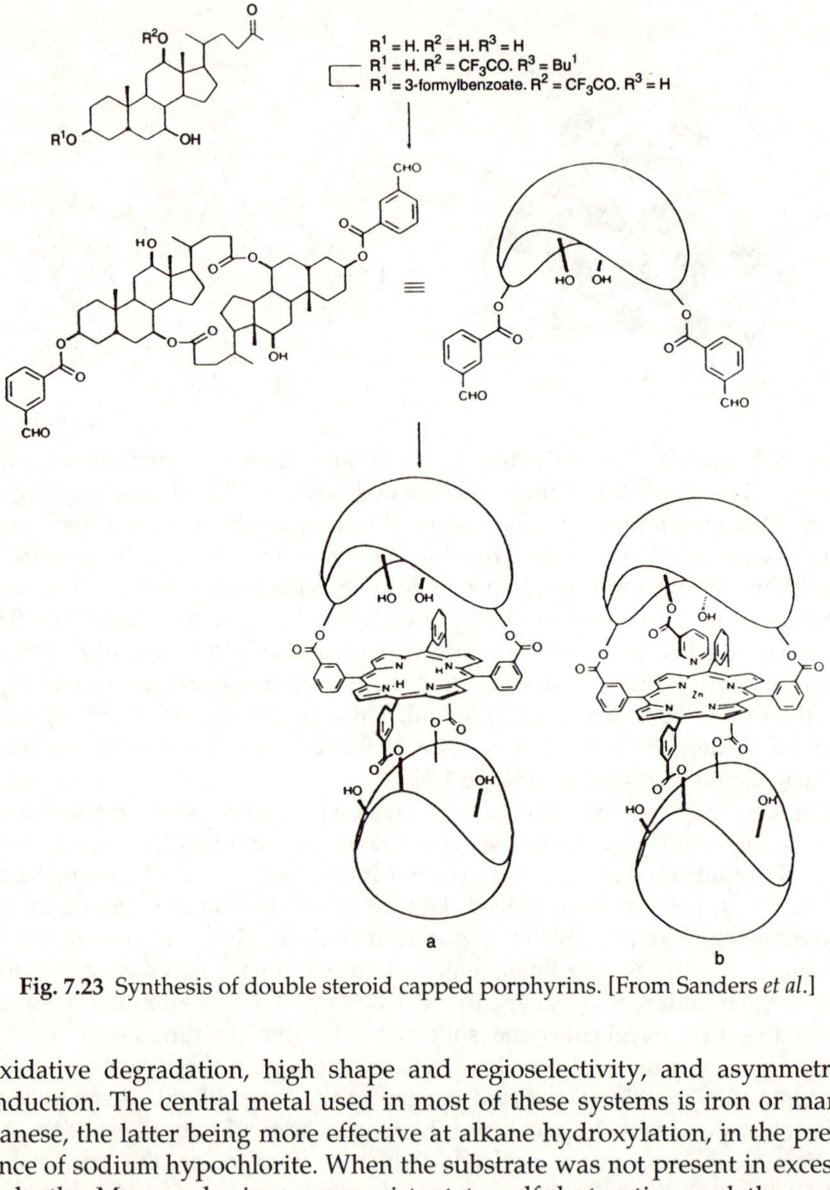

Fig. 7.23 Synthesis of double steroid capped porphyrins. [From Sanders *et al.*]

oxidative degradation, high shape and regioselectivity, and asymmetric induction. The central metal used in most of these systems is iron or manganese, the latter being more effective at alkane hydroxylation, in the presence of sodium hypochlorite. When the substrate was not present in excess, only the Mn porphyrins were resistant to self-destruction and the most effective ligand proved to be *meso* tetrakis(2,6-dichlorophenyl)porphyrin.[26]

Fig. 7.24 Synthetic porphyrins that model cytochrome P450s (a) protection of metal centre from oxidation; (b) shape and regioselectivity; and (c) asymmetric induction. [After Gunter and Turner.] In (c) the space-filling shapes represent binaphthyl units.

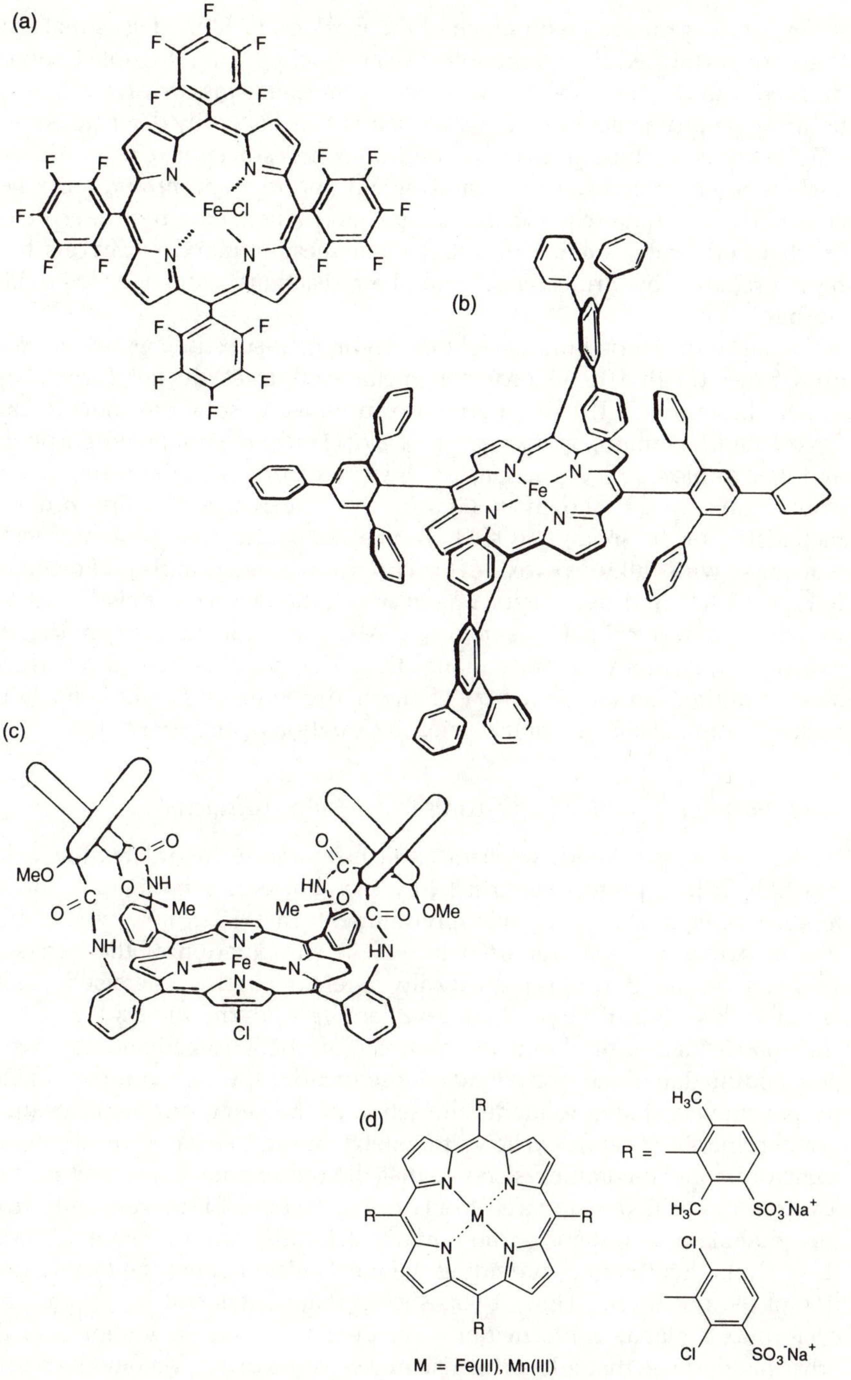

(a)

(b)

(c)

(d)

M = Fe(III), Mn(III)

R =

One of the problems with much of the work on P450 models is that reactions are performed in organic solvents in which it is not possible to obtain detailed information about the reaction mechanisms involved. This is because the proton activity in organic solvents is not easily determined. It is only in aqueous solution that the conditions necessary for oxygen transfer, such as ionic strength, acidity, and ligand species concentration, are best controlled, and data (e.g. electrochemical and kinetic) are best interpreted. To that end, water-soluble iron and manganese tetraarylporphyrins have been prepared by Bruice *et al.* and their reactions with hydroperoxides studies.[19b]

Owing to the steric hindrance of the eight *ortho*-substituents on the *meso*-artyl rings (methyl or chloro), the metal complexes do not form μ-oxo dimers [as most Fe(III) porphyrins do in aqueous solution], nor do they aggregate in solution, as does *meso*-tetrakis(4-sulphonatophenyl)porphyrin and its complexes. The reactions of alkyl peroxides with the water-soluble iron complexes are shown in Figure 7.25. The reaction is first order in metal(III) porphyrin and ROOH. Bruice concludes that oxygen-transfer reactions with alkyl peroxides occur from the complex (Porph)M^{III} (OH_2)(ROOH), and two others, one in which the ROOH is ionised and one in which a water ligand is ionised. Also, the rate of oxygen transfer increases upon stepwise proton ionisation. The model system also involves a rate-limiting homolytic fission of the hydroperoxide O—O bond (after preassociation of solvent water) prior to oxidation of any substrate.

7.4.4 Porphyrins with readily oxidisable meso-substituents

In the previous sections, we have seen how nature modulates the redox function of iron protoporphyrin IX by the choice of axial ligands and by modifications in the apoprotein environment. In 1983, a new type of porphyrin oxidation[27] was reported in which manipulation of the *meso*-substituents of the porphyrin drastically alters its redox chemistry. Thus, *meso*-tetrakis(3,5-di-t-butyl-4-hydroxyphenyl)porphyrin undergoes a two-electron oxidation in basified solutions of dichloromethane to give a porphodimethane-like compound. Subsequently, the iron complex of this porphyrin was shown to model the action of the heme prosthetic group in cytochrome P450, while cyclic voltammetry of the free base porphyrin and some of its metal complexes shows that the redox potential is significantly lowered in basified, compared to neutral, solutions. Electron density from the phenolate π-system is thought to delocalise on to the macrocycle through its deformation, providing the mechanism behind the facile oxidation of the porphyrin. Thus, it was shown that metals that lock the macrocycle into a planar conformation (e.g. Pt, Pd) impede oxidation. On the other hand, those that allow the macrocycle some distortion (and so permit

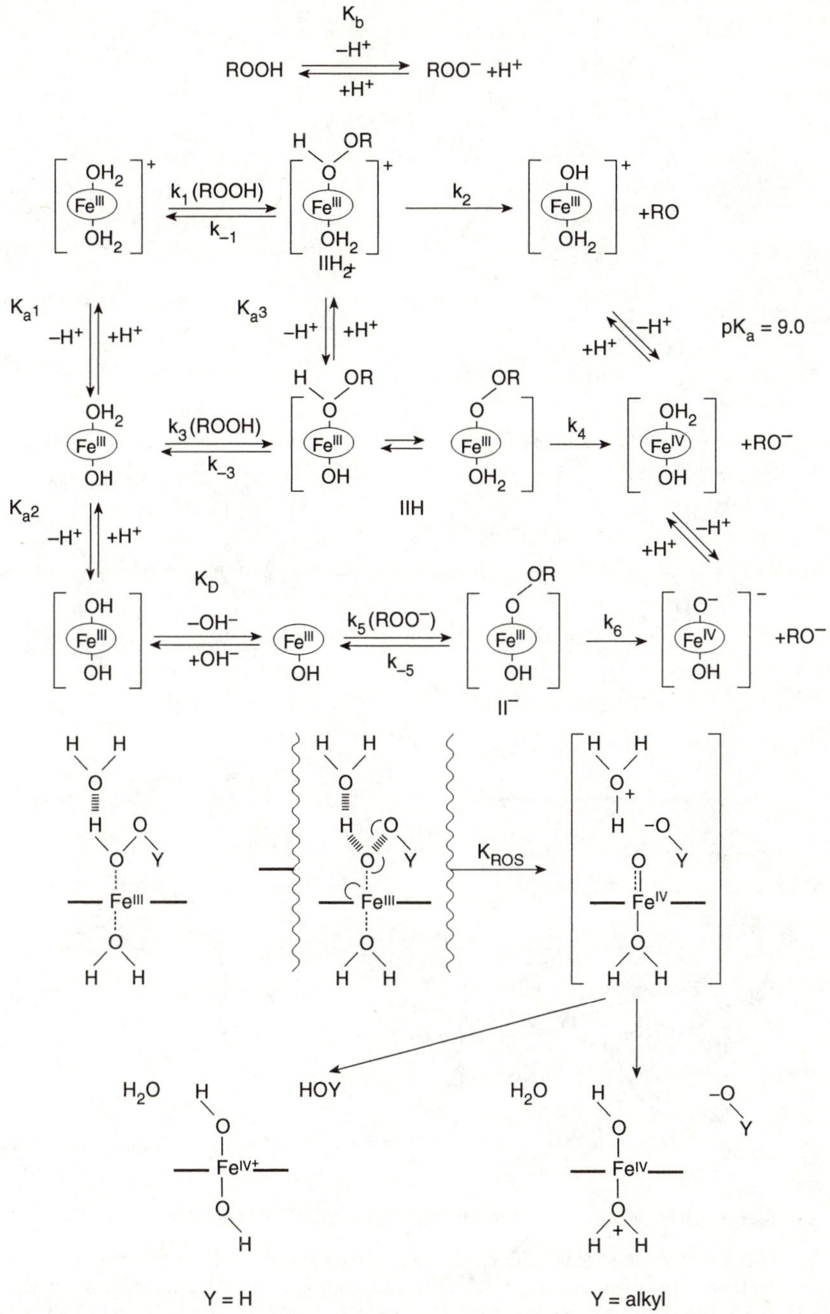

Fig. 7.25 Mechanism of reaction of water-soluble iron porphyrins with alkyl peroxides in aqueous media. [After C. Bruice, *Acc. Chem. Res.*, (1991), **24**, 243.]

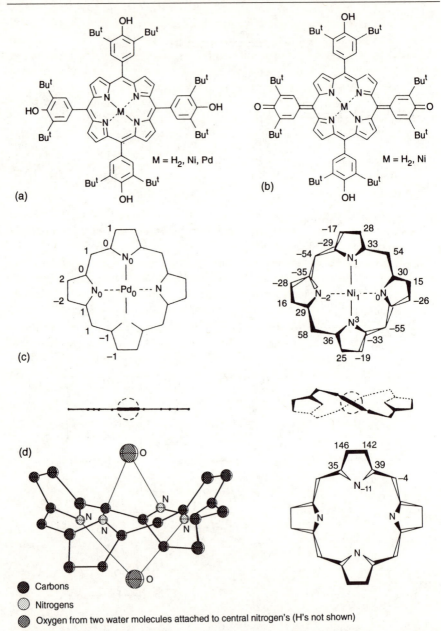

Fig. 7.26 (a) *Meso*-tetrakis (3,5-di-t-butyl-4-hydroxyphenyl)porphyrin and (b) its two-electron oxidation product. (c) Difference in macrocyclic distortion in the Pd and Ni complexes and (d) macrocyclic distortion of the oxidation product. [After Milgrom *et al.*]. The numbers indicate the degree of deviation from a mean plane through the molecule.

a certain amount of overlap between the phenolate and macrocycle π-systems, e.g. Ni) will undergo oxidation. Confirmation of this view was obtained from the crystal structure of the two-electron oxidised porphyrin which was shown to have one of the most severely distorted of tetrapyrrolic macrocycles with planar *meso*-substituents.

Addition of base (1M tetra-n-butylammonium hydroxide in methanol) to DCM solutions of the porphyrin gave a triplet ESR signal. This was interpreted as unpaired electron density localised on a *meso*-substituent. Initially, it was thought that the oxidation of the porphyrin proceeded stepwise with the direct reduction of oxygen to superoxide. Further studies, using spin-traps, however, indicate that the porphyrin undergoes nonstepwise two-electron oxidation, possibly reducing oxygen to peroxide, which then disproportionates to give hydroxyl radicals. The porphyrin radical is produced via a conproportionation reaction between the porphyrin and the two-electron oxidised porphyrin.

The porphyrin also undergoes aerial oxidation in acid solutions. This time, a π-cation radical is generated with unpaired electron density delocalised over the macrocycle. This radical is also produced via a conproportionation reaction between the porphyrin and the two-electron oxidised compound. There is some evidence that the π-cation radical, as generated from the porphyrin, is in fact a radical dimer. It is interesting to note that the aerial oxidation of the porphyrin affords radicals that can apparently be made to switch their unpaired electron density from different parts of the molecule depending on the acidity or basicity of the surrounding medium.

Substitution of other readily oxidisable groups into the *meso*-position should increase the ability of such a porphyrin to react with oxygen. Thus,

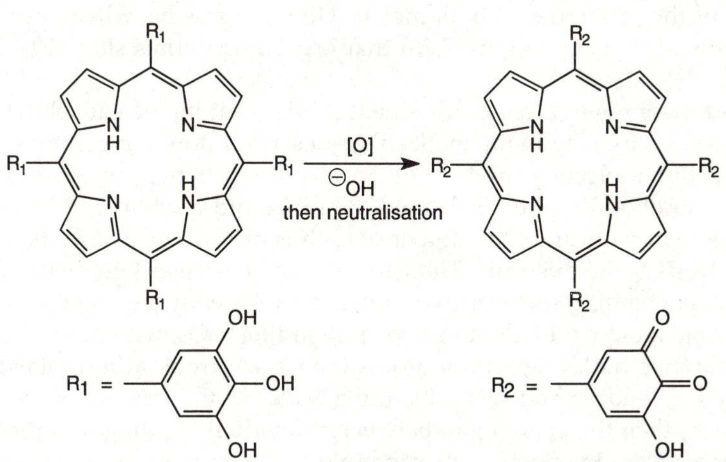

Fig. 7.27 Oxidation of *meso* tetrakis (pyrrogallyl) porphyrin to a tetrquinone.

on treatment with aqueous KOH, *meso*-tetrakis(pyrogallyl)porphyrin gives a deep-green ESR-active solution that turns brown-red as the spectrum decays. The spectrum is a triplet with side bands due to coupling to ^{13}C. Electrochemical studies using an oxygen electrode revealed that seven moles of oxygen per mole of porphyrin were required to generate the radical in two steps. On addition of concentrated HCl, a dark compound was precipitated which by microanalysis, UV–visible spectroscopy, and NMR suggested that a porphyrin tetraquinone had been formed. Oxidation of the tetraquinol porphyrin in total consumes eight moles of oxygen per mole of porphyrin. Presumably, the porphyrin molecule acts as an electron sink that couples the aerial oxidation of the *meso*-substituents.

Potential applications of easily oxidisable porphyrins are as sources of reduced oxygen species (e.g. peroxide, hydroxyl radicals) and as molecular conductors—one of the subjects of the next section.

7.5 Porphyrins and the electronics revolution

7.5.1 Molecular conductors

Electrical conductivity is normally associated with metals; most organic solids, such as polythene and PTFE being electrical insulators. Until relatively recently, anyone suggesting that organic or organometallic solids could exhibit the electrical, magnetic, and optical characteristics associated with bulk metals would have been considered a candidate for an asylum. Then, during the late 1970s, a new type of material appeared, the so-called "molecular metals", which are covalent materials (that may contain the one or two metal atoms covalently bound into an organometallic structure) with many of the properties of bulk metals. How can this be, when even simple band theory tells us (Figure 7.28) that organic materials should be insulators?[28]

Band theory is to three-dimensional solids what molecular orbital theory is to molecules. The mathematics that describes how we combine atomic orbitals into molecular orbitals (MOs), also predicts that for every bonding MO, an antibonding MO is formed. Bonding and antibonding MOs therefore come in pairs and the number of each is directly related to the number of electrons in the molecule. Thus, in a molecule such as butadiene, the four p orbitals combine to form two bonding MOs, with less energy than the individual atomic orbitals, and two antibonding MOs, with more energy. In a polyatomic molecule with n atoms (and n electrons in n orbitals), there will be $\frac{1}{2} n$ bonding and $\frac{1}{2} n$ antibonding MOs. As the number of these MOs increases, then the energy gap between individual bonding or antibonding MOs decreases. Eventually, we can build this picture up to describe a solid with an infinite three-dimensional array of atoms. Now, the number of

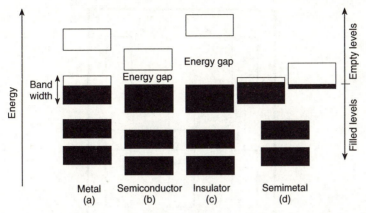

Fig. 7.28 Schematic depiction of electon occupancy of allowed energy bands for a classical metal, a semiconductor, an insulator, and a semimetal. The energy of the highest occupied level is called the Fermi energy. The unoccupied energy levels are white, the occupied levels black. [After Marks.]

bonding and antibonding orbitals is so great and the energy gap between them so small, that they merge into energy bands. The bonding MOs form a valence band (VB), while the antibonding MOs form a conduction band (CB). The gap between these two bands (called the band gap) is the property that determines a solid whether as a conductor, semiconductor, or insulator.

Electrons in the VB of a solid are localised around individual atoms in the lattice. If electrons manage to reach the CB, however, they can move freely through the lattice when an external electric field is applied, so that the solid is now electrically conducting. When the band gap is large, the solid is an electrical insulator. The energy needed to excite electrons into the CB is so large that before conduction can occur the lattice structure of the solid would break down.

In metallic conductors, on the other hand, the band gap does not exist. The electrons cross easily from the VB to the CB, where they can roam through the metal lattice at will. Metals may be considered to be atomic lattices awash in a "sea" of conducting electrons. Semiconductors have a band gap but, compared to insulators, it is narrow. Thus, a small amount of energy from, say, heat or light, will be enough to kick electrons up into the conduction band. Semiconductors also have two modes of conduction. Excited electrons move through the CB in negative, or "n-type", conduction, but where the electron has been excited from the VB, a positive "hole" is left behind. This can migrate from atom to atom in the lattice (which is another way of saying that VB electrons are hopping from atom to atom in the opposite direction, filling up the vacant space but creating another one

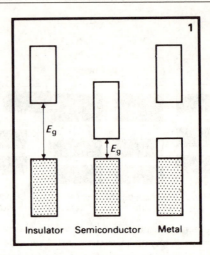

Fig. 7.29 Band theory for insulators, semiconductors, and metals. The shaded areas represent electronic states occupied by electrons. E_g is the energy gap between occupied and empty states. Metals have partly filled bands. Typically, E_g is over 4 eV for an insulator and below 2 eV for a semiconductor.

in turn). This positive hole hopping constitutes another form of conduction, called positive, or "p-type", conduction. Under the influence of an external field, the CB electrons move in one direction, while the VB "holes" move in the opposite direction.

In a pure semiconductor, the number of n- and p-type charge carriers exactly balances out. The conduction in such a semiconductor is said to be intrinsic. "Doping" with tiny amounts of impurities alters the number of charge carriers in favour of n- or p-type conduction, depending on the nature of the dopant. If it is a group III element, such as aluminium, boron, or gallium, then p-type conduction becomes paramount. The semiconductor is then called p-type. If the dopant is a group V element, such as phosphorus or arsenic, then n-type conduction is paramount and the semiconductor is called n-type. In both cases, the conductivity that arises from doping is said to be extrinsic.

The above discussion has been used to describe (highly qualitatively) the situation in inorganic semiconductors, such as doped silicon, gallium arsenide, etc. But the same arguments apply to organic materials, with one important difference. Whereas the conductivity of metals and inorganic semiconductors tends to be isotropic (i.e. the same in all dimensions), the conductivity of organic materials (when, of course, they are not insulating) tends to be highly anisotropic, i.e. it is associated with a preferred direction, by several orders of magnitude.[29] Thus, polyacetylene (Figure 7.30) in its virgin state is highly insulating. However, if an oxidising agent is added

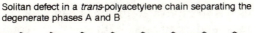

Solitan defect in a *trans*-polyacetylene chain separating the degenerate phases A and B

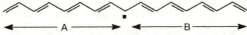

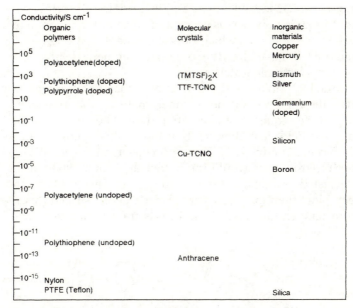

Room temperature conductivity values for various organic and inorganic materials. The units of conductivity are $\Omega^{-1}\text{cm}^{-1}$ or $S\,\text{cm}^{-1}$.

Fig. 7.30 Polyacetylene and its range of conductivities.

(for example, iodine, which removes electrons, giving the polymer p-type character), or a reducing agent (e.g. an alkali metal, which adds electrons giving an n-type polymer), then the polyacetylene becomes highly conducting. The change from an insulating to a conducting state involves an increase in conductivity of about 12 orders of magnitude. Also, the amount of dopant required to produce conductivity in an organic conductor can amount to a sizeable proportion of the weight of the polymer (*c.* 10%). This differs from the situation with inorganic semiconductors, where the amount of dopant required is usually in the parts per million range. With polyacetylene, this effect was first noticed in 1977. Since then, by clever manipulation of the doped polymer, it has been possible to increase the conductivity to such an extent that polyacetylene can now be produced which is as conducting as metallic copper.

This immediately gives some idea as to why research into conducting organic materials has so captivated the imagination of chemists, physicists, materials scientists, and engineers. With inorganic conducting materials, the conductivity arises as a result of the manufacturing process, which therefore has to be performed to exacting, highly expensive standards. With organic materials, the conductivity is a result of the molecular structure of the material, which, ideally, can be built-in prior to processing. Also, the processing of organic polymers is easier and cheaper than inorganic materials. Conducting polymers, of which polyacetylene is just one (see Figure 7.31 for some others), have already scored one commercial success, as batteries for powering electronic watches.

In conducting polymers, the conductivity may be envisaged as progressing along the direction of the polymer chain by the movement of stable "defects" in the sequence of alternating double bonds, under the influence of an applied field. Just alternation of the double bonds might seem to be enough to cause conductivity. This is, however, an oversimplification. Such a shift of electron pairs would imply that the ground state of the polymer chain consisted of a delocalised electron cloud. This has been shown to be incorrect. More than forty years ago, Peierls demonstrated that there was a minimum temperature (called the Peierls transition) below which such a

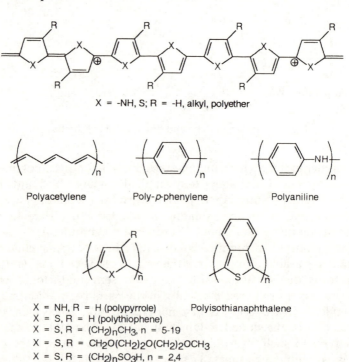

X = -NH, S; R = -H, alkyl, polyether

Polyacetylene   Poly-*p*-phenylene   Polyaniline

X = NH, R = H (polypyrrole)   Polyisothianaphthalene
X = S, R = H (polythiophene)
X = S, R = $(CH_2)_nCH_3$, n = 5-19
X = S, R = $CH_2O(CH_2)_2O(CH_2)_2OCH_3$
X = S, R = $(CH_2)_nSO_3H$, n = 2,4

Fig. 7.31 Other conducting one-dimensional polymers.

one-dimensional conductor could not sustain the long-range order necessary for a delocalised ground state to exist. Instead, lattice distortions would take over, leading to the conducting electrons being localised in a filled valence band which separates, energetically speaking, from an empty conduction band. This localisation of charge into regions of charge density could lead to conductivity if these regions were somehow coupled to the vibration of the lattice. It is the movement of the defects, rather than the motion of electrons under the influence of an applied field, that leads to conductivity in one-dimensional polymers.

There are other types of organic conducting materials. For example, there are organic charge-transfer complexes and ion radical salts, which are typically highly crystalline solids that consist of two component molecules. Thus, one molecule is an electron donor, while the other is an electron acceptor. Interaction between these two species produces a charge-transfer complex, if the individual component molecules are neutral to begin with, or ion radical salts if the individual components are ionic. A good example of a charge-transfer complex is the combination of the electron donor tetrathiafulvalene (TTF) and the acceptor, 7,7,8,8-tetracyano-p-quinodimethane (TCNQ).

In 1973, it was the production of a complex of these two molecules (with a 1:1 stoichiometry) that began the era of so-called molecular metals. TTF–TCNQ has a room temperature conductivity of about 500 Scm^{-1}, which rises to about 10000 S cm^{-1} on cooling. At about 50° K the Peierls transition occurs and the conductivity rapidly drops away. There are now many organic conductors of this type based on analogues of TTF and TCNQ.

Conductivity in these charge-transfer complexes arises out of segregation of the component donor and acceptor molecules into two separate stacks. This is what makes these conducting complexes different from ordinary insulating charge-transfer complexes, where the donor and acceptor groups simply alternate in non-segregated stacks. The close proximity of the individual molecules in the segregated stacks leads to overlap of their π-clouds (the donor and acceptor molecules are aromatic species), and the production of an energy band along the direction of the stack. Conductivity then arises if this band is combined with another band that is less than half-filled. This means that in such a conducting donor–acceptor complex where the stoichiometry is 1:1, the charge transfer has to be less than complete, requiring a delicate balance between the ionisation potential of the donor and the electron affinity of the acceptor. Thus, for TTF–TCNQ, the average charge that is transferred from the TTF donor to the TCNQ acceptor is 0.59 of an electron.

Using similar donor–acceptor complexes, organic superconductivity was discovered in 1980 at 1° K and 1200 atmospheres pressure for the complex

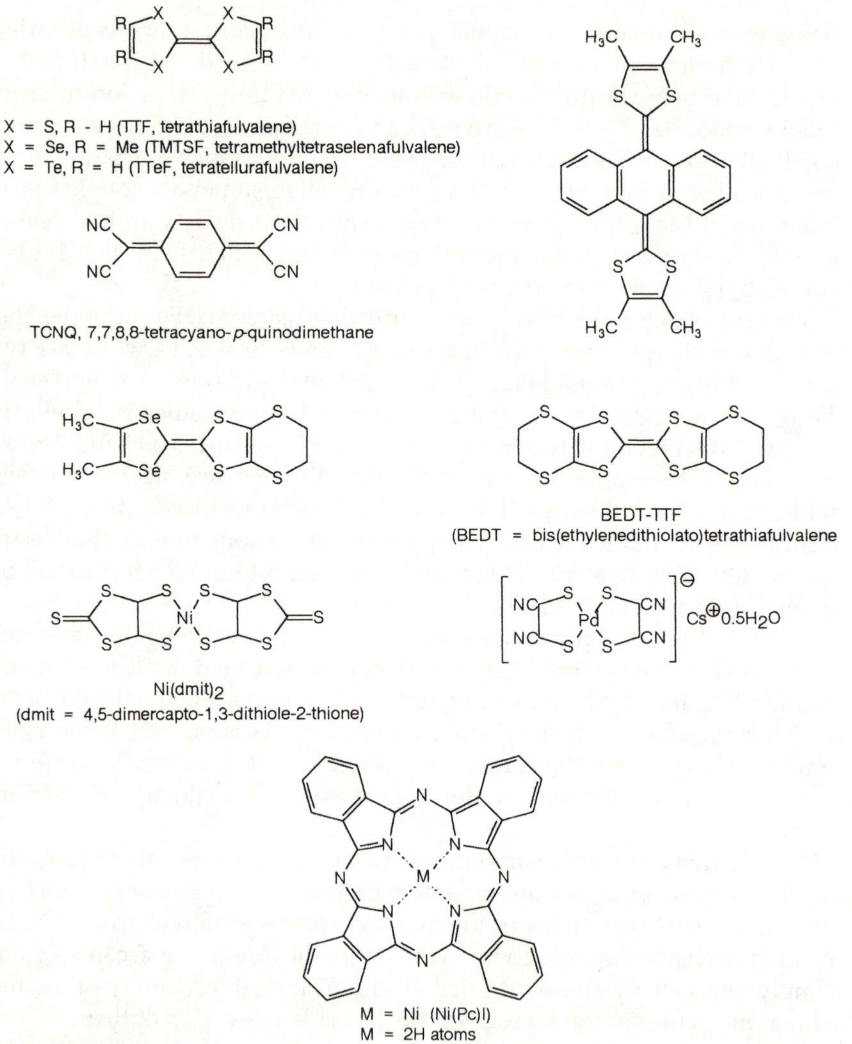

X = S, R - H (TTF, tetrathiafulvalene)
X = Se, R = Me (TMTSF, tetramethyltetraselenafulvalene)
X = Te, R = H (TTeF, tetratellurafulvalene)

TCNQ, 7,7,8,8-tetracyano-*p*-quinodimethane

BEDT-TTF
(BEDT = bis(ethylenedithiolato)tetrathiafulvalene)

Ni(dmit)₂
(dmit = 4,5-dimercapto-1,3-dithiole-2-thione)

M = Ni (Ni(Pc)I)
M = 2H atoms

Fig. 7.32 TTF–TCNQ, its analogues, and metallophthalocyanine two-dimensional conducting materials.

shown in Figure 7.30. the latter was needed to stop the Peierls distortion that normally occurs when the temperatures of these complexes are lowered far enough. By 1988, the temperature at which superconductivity appears had risen to 10.4 K with the copper thiocyanate salt shown in Figure 7.30.

Stacks of conducting metallomacrocycles (e.g. porphyrins and phthalocyanines) are ideal organic conductors, especially as p-type extrinsic semiconductors (i.e. the positive "holes" have to be added by doping, via oxidation with iodine, for example). Theory has predicted that the bare porphyrin macrocyclic nucleus, devoid of any *meso*- or β-substituents, should be a most promising conductor. The trouble is that of all the porphyrins, this is the most difficult to make in high yield and, when made, is highly insoluble and difficult to process.

Just as in any aromatic solid material, for there to be conductivity in a porphyrin and a phthalocyanine, there have to be small intermolecular stacking distances between the individual molecules to allow for extensive π–π overlap. Also, molecular subunits have to be in formal, non-integral oxidation states, which means that the highest energy band of the molecular stack is incompletely filled. As mentioned earlier, doping with iodine achieves this objective. Thus, in a series of nickel porphyrins and phthalocyanines, the conductivity increases by 12 orders of magnitude. Thus, conductivities of the order of 1500 S cm^{-1} have been measured. But probably the most valuable property of metallophthalocyanine conductors is that they remain conducting down to 1 K without the application of any external pressure. In fact, some of the most promising compounds are co-facial assemblies of group 4 (e.g. Si, Ge, and SN) phthalocyanines.[8]

Monomers of the group 4 metallophthalocyanines are prepared with two hydroxy groups coordinated to the metal on the fifth and sixth sites. Condensation polymerisation produces robust, highly crystalline, and rigid rod-like polymers, in which the number of subunits in the polymer (which are connected by μ-oxo bridges) typically varies from about 50 to 200. The polymers can be chemically or electrochemically doped. The conductivity

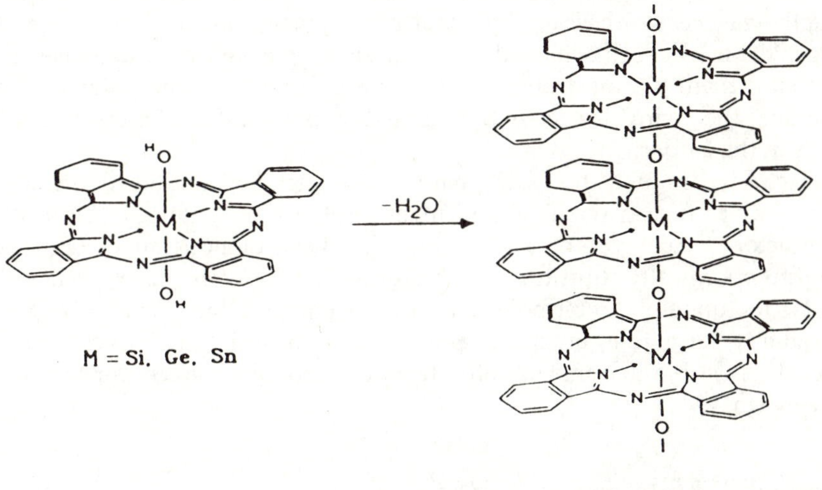

Fig. 7.33 Strategy for the co-facial assembly of structure-enforced group 4 metallophthalocyanine assemblies. [From Marks 1990.]

of these polymers is not as good as the crystalline monomeric charge-transfer metallophthalocyanines—about three to five orders of magnitude less—but what such polymers offer is a way to rationalise the design of organic conducting materials.

New classes of conducting polymers have been synthesised. For example, so-called "shish kebab" conducting polymers have been prepared using Fe, Ru, and Os porphyrins coordinated through the metal with bifunctional ligands, L–L, such as pyrazine and 4,4'-bipyridine. A development from this is porphyrin co-facial dimers (joined together by biphenyl or anthracene units) and coordinating second and third row transition metals capable of multiple bonding.[30]

In the metallomacrocyclic materials discussed so far, electrical conductivity is at right angles to the plane of the macrocycle. The synthesis of conducting stacks is difficult to perform in a controlled manner. It is this control that will be necessary if organic conductors are going to be useful in micro-, and eventually, nanoelectronic circuits, for example, as molecular wires, transistors, diodes, etc. Such nanoelectronic circuits will allow organic "chips" to be constructed where the number of circuit elements will not be counted in thousands but in billions. Computing power could be magnified by amounts hitherto undreamed of. There is a problem, however.

It is one thing to make molecules that can behave as miniature circuit elements. It is quite another matter to combine these molecules rationally into nanoelectronic circuits. Clearly, one cannot connect molecules together in the way that one solders wires! The molecules and their connections will have to be chemically synthesised. Thus, molecular wires of a desired length can be synthesised by connecting porphyrins together edge to edge.[31] This has been achieved by synthesising porphyrin tetraketones and coupling them via an aromatic tetramine to give a ladder polymer. The length of the "wire" is exactly determined by the number of monomer subunits in the ladder.

Electrons are not the only charges that can be conducted. Protonic conductivity is also possible, and many proteins are known to be proton conductors.[32] Such proton conductivity has been demonstrated with porphyrins specifically substituted with groups that allow the macrocyclic subunits to undergo intermolecular hydrogen bonding. The intriguing possibility exists that, by chemical manipulation of the macrocycle, it may soon be possible to achieve both electronic and protonic conduction in a porphyrin.

7.5.2 Porphyrins with liquid-crystal properties

Compounds whose molecules have a high degree of anisotropy, may be liquid crystalline.[33] When heating the solid, instead of melting to a clear

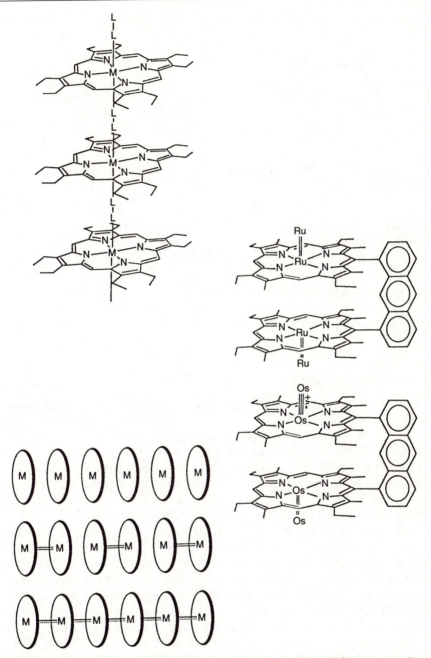

Fig. 7.34 (a) 'Shish-kebab' conducting polymers; L–L = 4,4'-bipyridine or pyrazine. (b) Porphyrin co-facial dimers and possible discotic array of metal–metal bonding causing conduction. [From Collman *et al.* 1990.]

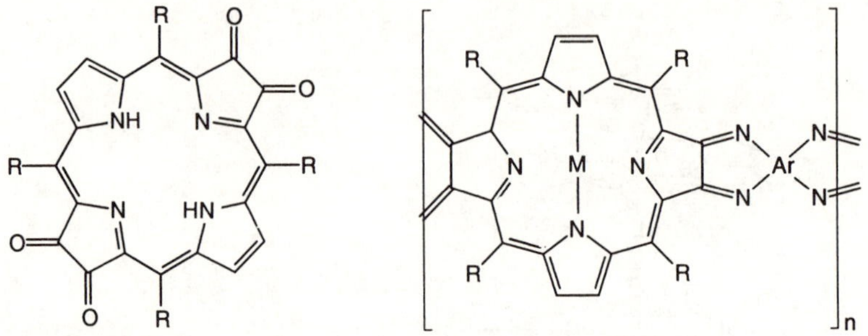

Fig. 7.35 Construction of edge to edge 'ladder' oligomeric porphyrins. Ar = an aromatic tetrammine.

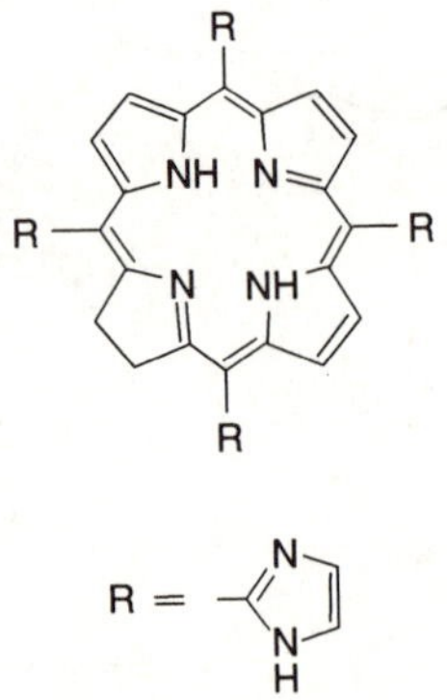

$$R = $$

Fig. 7.36 A proton-conducting imidazole-substituted porphyrin. [After Milgrom *et al.*]

isotropic liquid (in which the molecules are in random motion), a cloudy liquid is obtained (which remains between a fixed temperature range), in which there is some kind of time-averaged ordering of the molecules. Liquid crystals (LC) are liquids in so far as viscosity is concerned but resemble crystals in reflecting light and exhibiting birefringence.

The type of ordering within a liquid-crystal phase can be one-dimensional, in which the molecules point in an average direction (called the *director*) parallel with their long axis, but with little positional order in the layers. Such a phase is called nematic. If there is some positional orders in the layers (perpendicular to the director), the phase is called smectic A, whereas any other angle relative to the director gives a smectic C phase. Molecules within layers do diffuse into other layers, but they can be said to be spending more time on one or another layer and at different locations within the layers. This gives rise to many different types of smectic phases. Sometimes, one compound, on heating, can show several different liquid-crystal phases before finally giving a clear, isotropic liquid. Polarised light

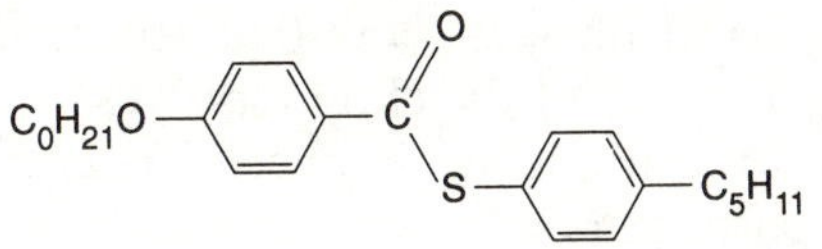

4-n-pentylbenzenethio-4'-n-decyloxy-benzoate

| 63°C–80°C | | Smectic A |

| 60°C–63°C | | Smectic C |

Fig. 7.37 Snapshot of molecules in two types of smectic liquid-crystal phases. [After Collings.]

will be transmitted or blocked depending on the orientation of the molecules in the liquid crystal. Liquid-crystal displays (LCDs), in watches and calculators, etc., depend on the ability to alter the molecular orientation with small electric fields.

Most of what has been aid so far applies to rod-like molecules, but similar phases exist for disk-shaped molecules, such as porphyrins and phthalocyanines. These tend to form phases in which the director is now orthogonal to the plane of the molecule. Thus, where the planes of the molecules are more or less parallel, but with no positional ordering into columns, the phase is a nematic discotic. If, however, the macrocycles form themselves into columns as well, we have a columnar smectic phase. A loose analogy of this phase is that it is like stacked coins, with the stacks arranged in a hexagonal lattice.

Phthalocyanines and metallophthalocyanines have been intensively studied as discotic liquid crystals,[34] but porphyrins have received less attention.[35] The first examples of porphyrin discotic phases were based on naturally occurring uroporphyrins but showed a very narrow discotic LC range. However, octa—substituted porphyrins can show discotic *meso*-phases over a broad temperature range, the LC phase being stabilised by metal insertion into the porphyrin, e.g. zinc. These materials are organic semiconductors whose LC properties allow them to form highly ordered thin films in the solid state. Emission, excitation and absorption spectra in the solid, LC, and liquid phases, show that observed spectral shifts are a function of the organisation of the chromophores, not just their proximity. Electrical conductivity in the discotic phase has also been observed in an octa——substituted porphyrin.

A recent observation has been that not all porphyrin LC phases are discotic. Thus 5,15-*meso*-substituted porphyrins have been prepared which show a variety of smectic phases that are not discotic.[36] The porphyrin macrocycle could be acting to impose biaxial symmetry within the *meso*-phase, which makes the new porphyrins of great interest as new nematic liquid-crystal materials.

It is becoming more and more apparent that conducting, processable, easily chemically modifiable and light weight organic materials offer the possibility of new technologies, as well as being important in fundamental solid-state chemical and physical research. Diverse applications, such as energy storage devices, sensors, switches, selective electrodes, "smart" windows, solar energy devices, non-linear optical materials, and electro-chromic displays have all been proposed and some are already being commercially exploited. Because of their unique blend of photoactivity and ability to transport electrons, porphyrins and related macrocyclic compounds are playing their part in these exciting developments. Truly, where there is life, there are porphyrins.

7.6 References

1. See A.H. Jackson, in *The Porphyrins*, Vol.1, ed. D. Dolphin, Academic Press, New York (1978), p. 374 and references therein.
2. G.T. Byrne, R.P. Linstead, and A.R. Lowe, *J. Chem. Soc.*, (1934), 1017.
3. M. Gouterman, *J. Mol. Spectrosc.*, (1961), **6**, 138.
4. L.R. Milgrom, *New Scientist*, (1984), **1395**, p. 26.
5. L.R. Milgrom, *The Independent*, 3rd September, 1990, p. 13.
6. J.R. Darwent, P. Douglas, A. Harriman, G. Porter, and M.-C. Richoux, *Coord. Chem. Rev.*, (1982), **44**, 833; M. Grätzel, *Acc. Chem. Res.*, (1981), **14**, 376; J. Kiwi, K. Kalyanasundaram, and M. Grätzel, *Struct. Bond.*, (1982), **49**, 37.
7. M.R. Wasielewski, *Chem. Rev.*, (1992), **92**, 435 and references therein.
8. M. Grätzel, in *Energy Resources Through Photochemistry and Catalysis*, ed. M. Grätzel, Academic Press, New York (1983), Ch. 3.
9. W.J. Albery, *Acc. Chem. Res.*, (1982), **15**, 142.
10. V.H. Houlding, K. Kalyanasundaram, M. Grätzel, and L.R. Milgrom, *J. Phys. Chem.*, (1983), **87**, 3175.
11. M.-C. Richoux, P. Neta, A. Harriman, S. Barel, and P. Hambright, *J. Phys. Chem.*, (1986), **90**, 2462; P. Neta, M.-C. Richoux, A. Harriman, and L.R. Milgrom, *J. Chem. Soc. Faraday Trans. 2*, (1986), **82**, 209.
12. (a) A. Harriman, G. Porter, and P. Walters, *J. Photochem.*, (1982), **19**, 183; (b) J. Davila, A. Harriman, M.-C. Richoux, and L.R. Milgrom, *J. Chem. Soc., Chem.*

Commun., (1987), 525; (c) H.K. Rao, D.O. Hall, N. Vlachopoulos, M. Grätzel, M.C.W. Evans, and M. Seibert, *J. Photochem. Photobiol., B. Bio.*, (1990), **5**, 379.

13. J. Fajer, *Chem. Ind.*, (1991), 869 and references therein.
14. J. Dalton and L.R. Milgrom, *J. Chem. Soc., Chem. Commun.*, (1979), 609.
15. C.A. Hunter, J.K.M. Saunders, G.S. Beddard, and S. Evans, *J. Chem. Soc., Chem. Commun.*, (1989), 1765.
16. L.R. Milgrom and F. O'Neill, in *The Chemistry of Natural Products*, 2nd edn, ed. R. Thomson, Blackie, (1993), p. 363 and references therein.
17. P.A. Adams, in *Peroxidases and Catalases in Biology*, Vol.2, ed. J. Everse, CRC Press, Boca Raton, p. 171 and references therein.
18. D. Lexa and M. Saveant, in *Proceedings of the International Symposium on Redox Mechanisms and the Importance of Interfacial Properties in Molecular Biology, 3rd Meeting*, eds. G. Dryhurst and N. Katsumi, Plenum, New York (1988), p. 1.
19. (a) H.L. Anderson, R.P. Bonar-Law, L.G. Mackray, S. Nicholson, and J.K.M. Sanders, in *Supramolecular Chemistry*, eds V. Balzani and L. De Cola, Kluwer Academic Publishers, Netherlands (1992), p. 359 and references therein; (b) T.C. Bruice, *Acc. Chem. Res.*, (1991), **24**, 243.
20. J.E. Baldwin, *Top. Curr. Chem.*, (1984), **121**, 181; B. Morgan and D. Dolphin, *Struct. Bond.*, (1987), **64**, 115.
21. C.A. Hunter, *et al.*, *J. Am. Chem. Soc.*, (1990), **112**, (a) p. 5773; (b) p. 5781; and (c) p. 5525.
22. (a) H.L. Anderson and J.K.M. Sanders, *Angew. Chem. Int. Edn.*, (1990), **29**, 1400; (b) H.L. Anderson and J.K.M. Sanders, *J. Chem. Soc., Chem. Commun.*, (1992), 946.
23. R.P. Bonar-Law and J.K.M. Sanders, *J. Chem. Soc., Chyem. Commun.*, (1991), 574.
24. M.J. Gunter and P. Turner, *Coord. Chem. Rev.*, (1991), **115**, 161.
25. T.L. Poulos, B.C. Finzel, and A.J. Howard, *J. Mol. Biol.*, (1987), **195**, 687.
26. S. Banfi, F. Montanari, and S. Quici, *Red. Trav. Chim., Pays-Bas.*, (1990), **109**, 117.
27. See reference 16, p. 347 and references therein.
28. See T.J. Marks, *Angew. Chem. Int. Edn.*, (1990), **29**, 857 and references therein.
29. See M.R. Bryce, *Chem. Brit.*, (1988), 782.
30. J.P. Collman, *et al.*, in *Organic Superconductivity*, eds V.Z. Kresnin and W.A. Little, Plenum Press, New York (1990), p. 359.
31. S.C. Narang and S. Ventura, Patent Application, US 4908442A, 13. 3. 1990.
32. L.R. Milgrom, S. Bone, D.W. Bruce, and M.P. Macdonald, *J. Mol. Electron.*, (1991), **7**, 95.
33. P.J. Collings, in *Liquid Crystals: Nature's Delicate Phase of Matter*, Adam Hilger, 1990.
34. J. Simon and C. Sirlin, *Pure Appl. Chem.*, (1989), **61**, 1625.
35. See reference 16, p. 374 and references therein.
36. Bruce, D.A. Dunmur, L.S. Sauta, and M.A. Wali, *J. Mater. Chem.*, (1992), **2**, 363.

INDEX